The Distribution of Subsistence Herring Eggs from Sitka Sound, Alaska

Sitka Sound core subsistence herring egg gathering areas with recent closures to commercial sac roe fishing

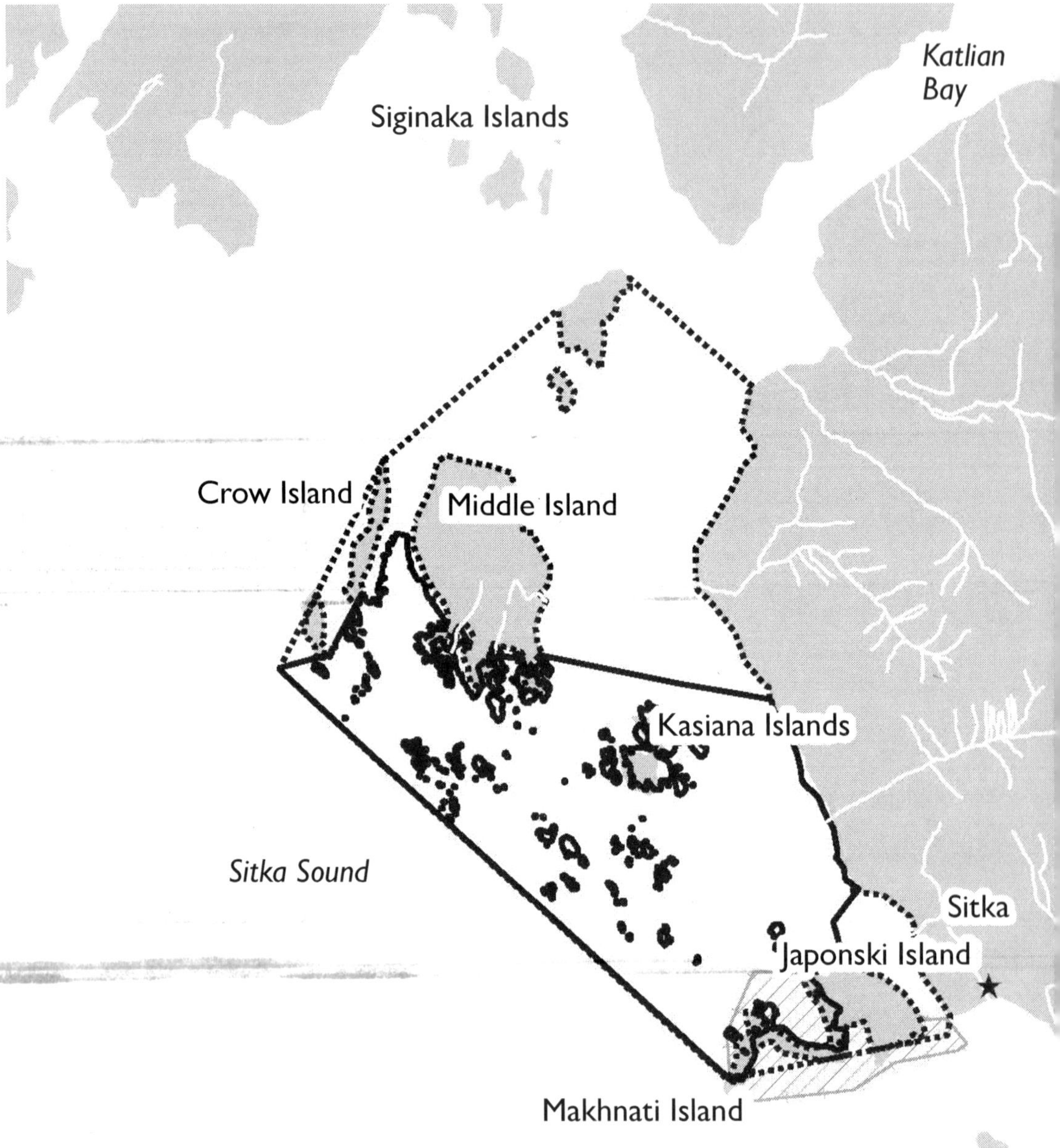

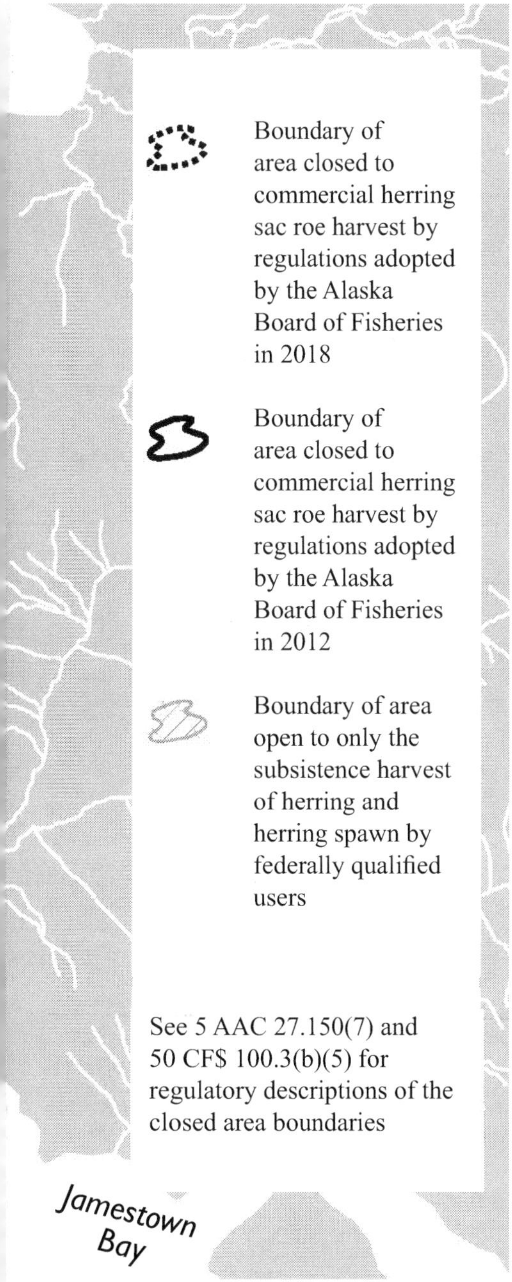

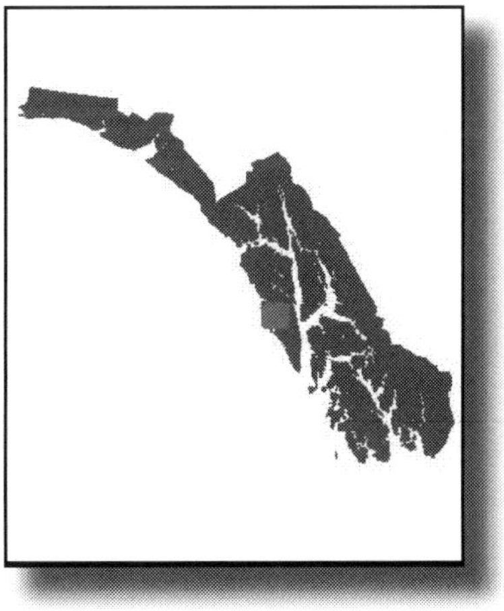

Above: Southeast Alaska with Sitka Sound denoted in red. Left: Sitka Sound and surrounding area.

The Distribution of Subsistence Herring Eggs from Sitka Sound, Alaska

By Thomas F. Thornton, Ph.D.

BOX OF KNOWLEDGE SERIES
SEALASKA HERITAGE INSTITUTE

Sealaska Heritage Institute is a private nonprofit founded in 1980 to perpetuate and enhance Tlingit, Haida, and Tsimshian cultures of Southeast Alaska. Its goal is to promote cultural diversity and cross-cultural understanding through public services and events. SHI also conducts social scientific and public policy research that promotes Alaska Native arts, cultures, history and education statewide. The institute is governed by a Board of Trustees and guided by a Council of Traditional Scholars, a Native Artist Committee, and a Southeast Regional Language Committee.

SHI's Box of Knowledge Series consists of essays, reports, and books that the institute considers should be made available as a contribution to studies on Tlingit, Haida, and Tsimshian cultures, history, and languages. They may be based on work carried out by researchers working in collaboration with SHI, contributions prepared by external experts, and work by staff. Publications in the Box of Knowledge Series will be available through SHI's website at www.sealaskaheritage.org and in book form.

SEALASKA HERITAGE INSTITUTE
105 S. Seward St., Suite 201
Juneau, Alaska 99801

Box of Knowledge graphic by Yukie Adams. Cover photo by Bethany Goodrich. Cover design by Nobu Koch. Book design by Kathy Dye.

ISBN 978–1–946019–43–1

SEALASKA HERITAGE

CONTENTS

List of Figures

List of Tables

FOREWORD

Dr. Tom Thornton is able to conclusively demonstrate the wide distribution and sharing of herring eggs and that it is a keystone species of the Tlingit, Haida, and Tsimshian cultures. Sitka harvesters generously shared more than 50 percent of their harvest with non-Sitka residents living throughout the state from Bethel, Kotzebue, and Utqiagvik (previously Barrow) from which they were further shared with outlying communities. Sharing of subsistence foods is a significant cultural practice that serves to unify kin groups and communities. Even with the dramatic decline in the availability of herring eggs, Natives have continued their ancient practice of sharing.

Based on their first-hand observations of the herring stocks, Dr. Thornton records that Natives have been telling the Alaska State Board of Fisheries for decades that the herring spawning populations in Sitka Sound have been diminishing as a result of the commercial herring sac roe fishery and the state's management regime. However, the state refuses to accept the existing scientific evidence and traditional ecological knowledge that they are managing a depleted resource. Neither do they accept that herring are a foundation forage fish for salmon, sea mammals, and other fish and wildlife in the marine food chain. Another clear management failure that Dr. Thornton's study demonstrates is that the State's allocated amount for Sitka herring eggs grossly underestimates the amount needed for subsistence use.

Unless the State of Alaska changes its management approach to the commercial herring fishery or establishes a moratorium on the commercial herring fishery, the ancient subsistence cultural tradition based on the harvest, distribution, and utilization of herring eggs will become extinct. We will see the demise of yet another Indigenous cultural practice. Social cohesion will be undermined with decreasing numbers of harvesters gathering and sharing herring eggs. Traditional knowledge about herring behavior, migration patterns, and differences between male and female herring will be lost to succeeding generations.

Dr. Thornton's study is one of the most extensive studies of its kind on the cultural, historical, and societal importance of herring in Sitka Sound, the mecca of herring spawning and egg gathering for hundreds if not thousands of years and the last stronghold in the region of this keystone species. We can hope that this study will not be a record of a cultural tradition that has succumbed to poor fisheries management and a commercial harvest.

—Rosita K̲aaháni Worl, Ph.D.
President, Sealaska Heritage Institute

ACKNOWLEDGEMENTS

I am very grateful to Sealaska Heritage Institute, and especially Chuck Smythe and Rosita Worl, for conceiving and supporting this study of the annual distribution of herring eggs harvested from Sitka Sound, which occurs each spring. I also am indebted to members of the Sitka Tribe of Alaska, especially Kathy Hope Erickson, Lisa Gassman, Jeff Feldpausch, Harvey Kitka, Mike Miller, Kyle Rosedale, and Tammy Young, and the Hoonah Indian Association, especially Robert Starbard and David See, who assisted me with coordinating and carrying out the research in their respective communities. I also wish to thank Forest Haven for her help with interviews in Sitka and Metlakatla; Lauren Sill of the Alaska Department of Fish and Game, Division of Subsistence, for her help with the state data and integrating the supplemental questionnaire into the 2018 herring egg harvest survey; Mason Auger, Kathy Dye, Amy Fletcher, Will Geiger, and Nobu Koch of SHI for their assistance with manuscript review, layout, and production; Bethany Sonsini Goodrich for her wonderful photographs; Adam Kersch for select interview transcriptions; and Jamie Hebert for her invaluable work in transcribing interviews and assistance with the manuscript preparation. Most of all, I give thanks to all the herring egg gatherers, givers, and receivers, who graciously shared their time and expertise with me over the course of this study and who work tirelessly to maintain the magnificent distribution system that delivers herrings eggs and a host of co-benefits to relatives and friends in communities around the state and beyond. You care passionately about the health of Pacific herring because you see how its status is inextricably interwoven with your own culture and wellbeing.

Elijah Lawrence of the T'akdeintaan Tlingit clan displays Sitka Sound herring eggs on branches harvested by his family. (Photo: Bethany Sonsini Goodrich, Sustainable Southeast Partnership)

EXECUTIVE SUMMARY

This study describes and analyzes the distribution of herring eggs from Sitka Sound, based on primary and secondary research of historical and contemporary Native herring fisheries, including participant observation in Sitka and interviews with more than 50 participants in the distribution system in Sitka, Angoon, Hoonah, Kake, Juneau, Metlakatla, and other communities. The need for the study is based on current stress experienced by subsistence harvester-distributors and recipients of herring eggs, due to declining abundance and reliability of quality herring spawn in Sitka Sound.

Pacific herring (*Clupea pallasi* or *yaaw* in Tlingit) are a foundation forage fish, cultural keystone resource, and bellwether species for marine ecosystems in the Gulf of Alaska. Herring's status as a cultural keystone species is exemplified by the highly developed cultivation, production, and distribution system that residents of Sitka and other Native communities in Southeast Alaska developed to optimize the benefits of herring spawn. Herring eggs are gathered on broad (*daaw*) and hair (*né*) kelp and on Western hemlock branches (*haaw da.aa*), the latter being carefully laid out in sets to extend and enhance spawning habitat and thus the reliability, stability, and quality of herring egg harvesting opportunities in prime areas. This system was highly developed in Sitka Sound, the mecca of herring spawning and egg gathering for hundreds, if not thousands, of years. Archaeological evidence documents a 10,000-year tradition of herring harvest in the northern Northwest Coast region. Although the origins of the herring egg cultivation system are difficult to pinpoint in the archaeological record, oral histories of the Kiks.ádi Tlingit clan assert its deep antiquity. Herring egg cultivation originated with the discovery by a Kiks.ádi woman that herring at *Yaaw Teiyí* (Herring Rock), located in front of the Sitka Indian village, could be coaxed to spawn on introduced substrate in the form of the woman's hair. This discovery, combined with further experiments and learning, led to the development of a sophisticated knowledge-practice-belief complex and sustainability covenant with herring that has important material, social, and spiritual dimensions. The Herring Rock seascape, crest, story, and names derived from these events are all considered *at.óow* (sacred property) of the Kiks.ádi.

Today, on average, 87 percent of the 71,936-381,226 pounds of herring eggs harvested at Sitka Sound are distributed to relatives, friends, and other individuals and institutions through sharing, barter, and trade networks. This constitutes the highest distribution of any subsistence resource in Southeast Alaska. The generosity and exchanges underlying this massive distribution speak to the cultural keystone status of herring eggs as a "first fish," food security resource, symbol of identity and renewal, and its value in promoting social networking, social capital, and conviviality. The customary and traditional gift, barter, and trade economy characterizes herring distribution, rather than the market economy. Between 2002-2018, this subsistence economy delivered herring eggs from Sitka Sound to 41 communities throughout Alaska and beyond. Among those receiving the most eggs were Sitka and its neighboring villages of Angoon, Hoonah, and Kake, Also, receiving medium to high herring egg distributions were regional centers such as Juneau, Ketchikan, and Anchorage, as well as Washington State.

Since the 19th century, the Sitka Sound subsistence herring egg harvest has been severely affected by colonialism, habitat destruction (including the dynamiting of Herring Rock), and overfishing, especially with the rise of commercial herring reduction and sac roe fisheries in the twentieth century. The sac roe seine fishery has become particularly concentrated in Sitka Sound since the 1980s, wherein about 50 seine permit holders net up to 20,000 tons of herring just prior to spawning, typically in areas proximal to traditional subsistence gathering. These impacts have meant reduced opportunities for subsistence production and distribution of herring eggs. Even more devastating impacts from commercial fishing and development on other spawning areas (e.g., Auke Bay, Juneau) have literally wiped out spawning herring populations, thus increasing dependence on Sitka Sound as the last major source of herring eggs in Southeast Alaska. This double bind of reduced opportunity and increased dependency has put particular stress on contemporary Sitka Sound herring egg harvesters, who expend up to 10 times as much effort as they did formerly to obtain the same amount of quality eggs.

The successful harvest of herring eggs relies on sociocultural factors (knowledge, skills, time, and means) and ecological ones (sufficiency, quality, duration, reliability and accessibility of spawn). The current management regime has led to reduced spawning extensity, duration, predictability, and quality, in turn leading to increased competition and material and sociocultural costs to find and capture suitable spawn, either on Western hemlock branches (placed as substrate) or kelp (naturally occurring hair or macrocystis kelp). Areas once proximal and reliable are no longer supporting herring spawn or producing an unsatisfactory quality of spawn for subsistence uses. With fewer areas of spawn there also is increased jostling among harvesters, including from the commercial sac roe industry-supported boats the crews of which are paid to harvest eggs. These paid harvesters, who may take more than 40 percent of the total subsistence harvest, are further crowding out customary and traditional subsistence users.

The modern era has produced new technical means to harvest (e.g., motorized boats), preserve (e.g., freezing), and distribute (e.g., by air freight) herring eggs, which necessitate cash. The role of cash in the herring egg economy comes both through contributions to the production of herring eggs (e.g., donations of fuel money to lay hemlock branches) and through lawful customary trade (for cash) in their subsistence distribution. Commercial sales of herring eggs are illegal and comparatively rare, as there is strong social opprobrium towards those who do so, especially in rural Native communities.

New organizational means also have evolved to distribute herring eggs within and beyond Sitka. Traditionally, the herring egg harvest included many non-local Natives who came from other communities to participate in the harvest, process the eggs, and then transport and distribute them back home or to other communities. Similarly, local harvesters engaged not only in local distribution but also in gifting, barter, and trade missions to dozens of communities throughout Southeast

Alaska and as far away as interior Alaska, Northwest Canada, and the Western United States. Herring eggs were often traded for non-local delicacies such as caribou, muktuk, moose, eulachon, soapberries, and red seaweed; these barter exchanges continue today alongside customary trade for cash. More recently, individual and household distribution has been supplemented by organizational distributions of eggs, including through tribes (Sitka Tribe of Alaska and Hoonah Indian Association), Alaska Native Brotherhood camps, senior centers and hospitals, the sac-roe industry's Southeast Herring Conservation Alliance, and other charities, such as the Salvation Army. Local Sitka households and organizations also routinely assist non-local harvesters in obtaining herring eggs, despite the increasing competition over accessible spawning areas.

The benefits that flow from the distribution of Sitka Sound herring eggs are profound. Herring eggs support physical health and wellbeing; nutrition and food diversity; social connectivity; cultural heritage, identity, and transmission; and ceremonial, spiritual, and other cultural events. Perhaps more than any other single subsistence food, the distribution of herring eggs supports the values, institutions, and skills that define "the good life" in Tlingit and other Southeast Native communities.

This report finds that state management of the herring fisheries focuses too narrowly on maximizing yield for commercial sac roe at the expense and undervaluing of herrings' ecosystem service values, and especially the system of *cultivated abundance* underlying subsistence production and distribution. Subsistence herring eggs directly benefit thousands more Alaskans than the export sac roe fishery and its by-products. Additionally, suggestions for further study are offered, as are recommendations for management reform to ensure sufficient protection for the sustainment and restoration of the customary and traditional Sitka Sound herring egg fishery and the many values and benefits its remarkable distribution system supports. In sum, the report suggests the need to reform the current Maximum Sustainable Yield (MSY) model of commercial herring management in order to effectively support the Maximum Sustainable Benefits (MSB) of herring eggs delivered by the Alaska Native cultural economy.

CHAPTER 1: INTRODUCTION

SHARING AND DISTRIBUTION OF THE CUSTOMARY AND TRADITIONAL SITKA SOUND HERRING EGG HARVEST

"If you receive herring eggs from someone, you know you're loved."

—Vivian Mork, Sitka

Wade Martin packs a 50-pound "wetlock" box of herring eggs for shipment on Alaska Airlines to friends and relatives in other communities. On average, 87 percent of the herring egg harvest is given away through sharing, gifting, barter, or customary trade. (Photo: Bethany Sonsini Goodrich, Sustainable Southeast Partnership)

Perhaps the most remarkable thing about the annual Sitka Sound Pacific herring (*Clupea pallasi; yaaw*) subsistence egg harvest is how much of it is given away. In the 2002 to 2014 period, during the first 12 years of the Alaska Department of Fish and Game-Sitka Tribe of Alaska joint monitoring and survey program, the amount of the herring eggs harvested annually ranged between an estimated 71,936 and 381,226 pounds. This yields an annual average of 160,216 pounds of herring eggs, taken predominantly by local Southeast Alaska Native Tlingit, Haida, and Tsimshian peoples, or other members of the Sitka Tribe of Alaska (STA). The amount of this harvest that was given away—i.e., shared or exchanged—averaged 87 percent of the take, or approximately 140,000 pounds per year (Appendix 1). If we convert these figures into tons, an average of 70 tons of an 80-ton total harvest are given away, that is, distributed and redistributed throughout the State of Alaska—from Southeast's southernmost community of Metlakatla to the state's northernmost town of Utqiagvik (Barrow) and beyond. No other subsistence resource matches herring eggs in terms of the prodigious amounts given away in relation to the total harvest or in its expansive distribution range in relation to the total harvest area, which is presently confined almost exclusively to Sitka Sound. The distribution of herring eggs thus represents an extraordinary largesse, though not all herring eggs are given or shared as simple gifts.

This overview of the nature of the herring egg distribution system as sharing, barter, and trade must be understood against the broader characteristics and evolution of the herring fishery. For millennia thousands of Southeast Alaska Natives have fished herring as part of their seasonal rounds of food production—cooking and smoking the meat, rendering oil from the flesh, and harvesting eggs (*gáax'w*) during spring spawning on broad macrocystis (*daaw*) and hair kelp (*né*) and introduced Western hemlock boughs (*haaw da.aa*) substrates. The Tlingit, Haida, and Tsimshian peoples of Southeast Alaska were intimately familiar with herring life cycles, especially the spawning stage, when herring are nearshore and could be monitored closely. This is evidenced in Tlingit oral histories, where spawning grounds, such as Herring Rock (*Yaaw Teiyí*) for the Kiks.ádi, were discovered, named, and made sacred as result of their importance as spawning habitat and loci for Native egg harvesting. This Native relationship with herring was a covenant and involved cultivating the fish to return to their favored spawning grounds through respectful engagement and "relational sustainability" (Langdon 2007), which included a range of techniques described later in this report (see also Thornton 2015).

Distribution through Sharing, Barter, and Customary Trade

Under law, herring eggs are shared, bartered, or traded, the latter including what has become known as "customary trade." Sharing refers to the distribution of subsistence foods without receiving something immediately in return, though, as with a gift (Mauss 1967), there may be a generalized expectation of some return gift, service, or gesture in the future. Barter involves limited exchange of subsistence fish, wildlife, or other wild, renewable resources for goods other than cash. "Customary trade" is a term for exchange of subsistence fish, wildlife, or other wild, renewable resources for cash (Wolfe et al 2000).

Customary trade is the most controversial and circumscribed of these three types of transactions because it involves cash, thus blurring the boundary between subsistence and commercial activities and potentially desacralizing the more traditional sharing and barter forms of exchange, which have deeply embedded cultural roots. In discussions leading to the 1980 subsistence legislation known as ANILCA (Alaska National Interest Lands Conservation Act), the U.S. Senate Committee on Energy and Natural Resources stated categorically that,

> The Committee does not intend that "customary trade" be construed to permit the establishment of significant commercial enterprises under the guise of "subsistence uses." The Committee expects the Secretary and the State [of Alaska] to closely monitor "customary trade" component of the definition and promulgate regulations consistent with the subsistence title. (Senate Report No. 413, 96th Congress, 2nd Session, 234; see Schroeder and Kookesh 1990:37)

Customary trade, then, is not a profit-making scheme, but rather a legal right under subsistence law to trade non-commercial or subsistence amounts of herring eggs for cash. Trade for cash is recognized as a necessity in rural communities, both because it is customary and traditional (i.e., a practice going back generations) and because all Alaska communities, even remote ones, have long been "mixed economies" (Wolfe 1984; Wheeler and Thornton 2005) dependent on at least some cash inputs to purchase such things as fuel and equipment (boats, gear, etc.) to carry out subsistence activities. The introduction of cash in itself, however, did not transform the customary and traditional character of kin-based production and exchange in Alaska Native communities. Rather, these communities remain subsistence-based economies, incorporating cash as needed into existing production and value chains. This is because subsistence is predicated not simply on material transactions but also on a set of cultural values, which inform Alaska Native identity (Langdon and Worl 1981; Sahlins 1976). These unique Indigenous cultural values are recognized in ANILCA, which emphasized that: "The opportunity for subsistence uses…is essential to Native physical, economic, traditional, and cultural existence."

These values inform herring egg harvests in Sitka Sound, which are typically pursued by deploying Western hemlock tree boughs from motorized boats in island areas beyond Sitka's immediate shores or by using small boats to collect herring spawn on kelp (SOK) in select patches. Notwithstanding the necessary flow of some cash in the subsistence economy, under state law (as opposed to ANILCA and federal regulations), customary trade of subsistence resources for cash is prohibited, unless specifically authorized by the Alaska Board of Fisheries. Thus, under state law, the boundaries between commercial and subsistence fishing are maintained more strictly than those of non-fish subsistence resources (Thornton and Mamontova 2017).

Significantly, the only customary trade of fish legalized by state regulation in Southeast Alaska are herring eggs on kelp. This is a reflection of the importance of SOK in customary barter and trade (Langdon 1991), the former of which dates back millennia and the latter, trade for cash, to at least the late 19th century, when Haidas began selling SOK to non-Native buyers—typically Japanese or other Far Eastern interests—in Seattle and elsewhere (Victor-Howe 1992). A case brought against Haida herring egg sellers (938 F.2d 942 (9th Cir. 1991)) for attempting to sell thousands of pounds of herring eggs on kelp to international buyers was not successfully prosecuted. The court ruled that the quantities of eggs harvested were not clearly beyond those authorized under ANILCA's broad definitions of customary trade, including trade for cash (Schroeder and Kookesh 1990; Victor-Howe 1992:70). While this seemed at first to be a victory for subsistence users, the case resulted ultimately in the passing of new State of Alaska regulations to limit the customary trade of fish in Southeast Alaska exclusively to herring eggs. Thus, anyone seeking to sell fish other than herring eggs for cash may be subject to prosecution.

State and federal regulations (50 CFR 100.27(c) and 36 CFR 242.27(c)) introduced following the launch of the federal government's subsistence policy under Title VIII of ANILCA now govern the harvesting, processing, and sale of subsistence foods. These regulations: 1) provide for sales involving small amounts of fish and other renewable subsistence resources; 2) prohibit sale of subsistence fish to (and purchase by) businesses; 3) require non-rural purchasers of subsistence fish (e.g., a resident of Juneau, Anchorage, or Fairbanks) to use it for their own or their social network's own personal consumption (i.e., non-rural purchasers cannot resell fish); and 4) require uses be consistent with Federal Subsistence Regional Advisory Council policy and the intent of Title VIII of ANILCA. Figure 1 provides an overview of how these exchange relationships with fish are classified within the broader legal framework of subsistence and non-subsistence resource exchanges in Alaska.

Sitka Sound herring egg harvesters transfer herring eggs by air freight or other means to more than 50 communities around Alaska and beyond. The majority of these distributions are gifts to friends and relatives as a seasonal greeting and reminder of this special "first fruit," marking the beginning of the productive season for wild foods. Another portion is bartered, especially for other unique place-based foods or other goods, such as moose or eulachon, from the mainland communities. Trade between island and mainland Southeast Native communities has always been vigorous and significant in part because of the differences in geographies, microclimates, and ecological resources. Although historically herring could be found in quantity in proximity to many Native communities, significant populations of spawning herring were more limited, and since the late twentieth century spawning areas have become highly circumscribed due to overfishing for reduction and sac roe. Today, herring egg harvesting is confined to island Tlingit communities, especially Sitka and Klawock, which currently lie within the only active commercial herring roe fishing management areas, known as Sitka Sound and Craig (Frontispiece).

It must be acknowledged from the outset that it is difficult to parse precisely what percentage of the average 70-ton subsistence herring egg harvest is given away, bartered, or traded for cash. This is due in part to a lack of comprehensive research and reporting on each type of distribution, but it is also a function of the complexity of the herring egg economy, which defies this simple tripartite typology of transactions. A few hypothetical examples can be illustrative of the issues and preview the more in-depth analysis of the remaining chapters. First, consider how cash given to support herring egg production is not understood as cash "in exchange for" eggs. If Elder A lives in Juneau (where subsistence herring eggs can no longer be gathered due to the collapse of the fishery with the last commercial herring sac roe harvest in 1982) but has Nephew Harvester B who lives in Sitka (where herring eggs are accessible), Elder A may give the nephew (who may be lacking cash) fuel money so he can run a boat to carry out herring egg gathering. Nephew Harvester B, perhaps because his production capacity has been enabled or enhanced by the fuel money, but also because he loves and respects his uncle (or aunt), may send herring eggs back to Juneau for his Elder(s) to enjoy, perhaps paying the freight with left over fuel money, etc.

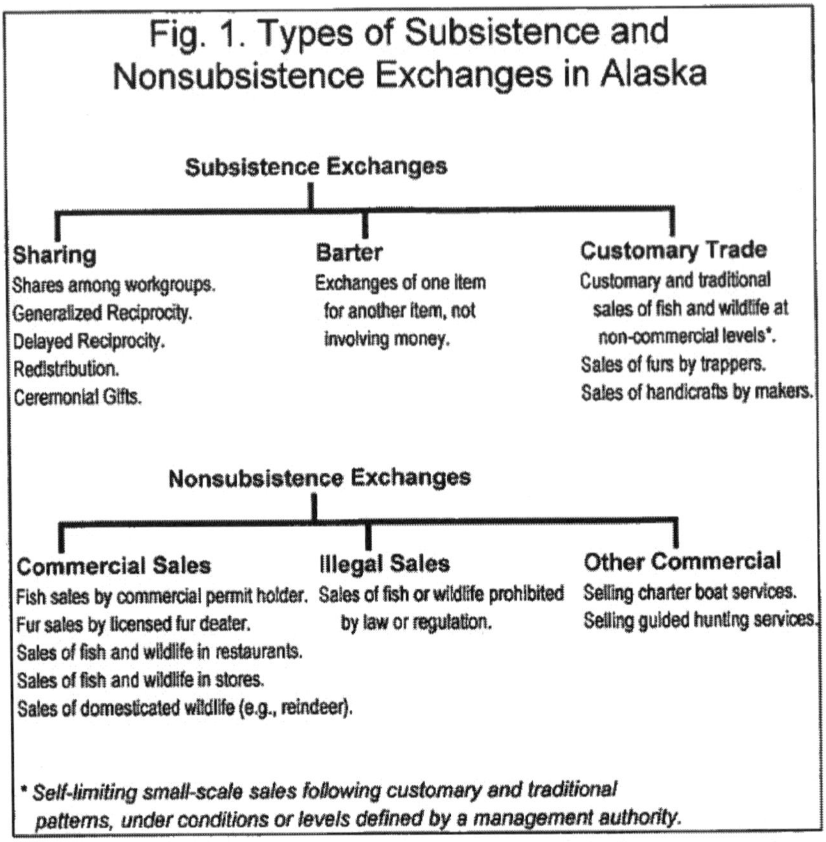

Figure 1. Types of subsistence exchanges (from Wolfe, et al 2000:3).

Alternatively, if Nephew Harvester B fails to collect enough eggs to send to Elder A, he would not return the fuel money, nor would his Elder expect him to do so. This is because neither regards the donation of fuel money as "trade for cash" but rather views it as a group orientation to production, where both parties—Elder and Nephew—are part of the same production unit. Elder A may support his nephew in production with cash because Elder A is not able to help directly with his own labor or fuel. He does not expect that he will get a certain amount of herring eggs in return for his investment in production. Elder A and Nephew Harvester B are in fact working together and both bear the risks of failure if the harvest of herring eggs fails. This is a very different orientation than that of a simple trade transaction, the logic of which might be articulated in this way: "Nephew, I will trade you $100 for a 50-pound box of herring eggs because that is a fair deal for that amount and the labor required to produce it." Such basic transactional logic is in fact uncommon in subsistence herring egg distribution—almost unheard of among relatives, but also not common among friends. In these more complex exchanges, there are nearly always other social objectives to be met beyond simple trade, from showing respect, to returning a favor, to cultivating new or revitalizing old relationships. The literature on subsistence is consistent on this point, as well as on the point that market exchange or outside trade (i.e., to other communities) does not attenuate subsistence ties within communities (cf. Langdon and Worl 1981).

In other cases, customary trade for cash is based at least partly on offsetting expenses. This may be negotiated, as in the following hypothetical example: Anchorage Recipient R, an acquaintance or "friend of a friend" of Sitka subsistence herring egg Harvester D, makes contact and asks, "Do you have any herring eggs down there you can send to me?" The harvester might respond, "Yes, I can give you a small box, but shipping has gotten expensive, so it would be good if you can help me out on that." Such transactions clearly do not constitute commercial exchanges. Many harvesters keep "lists" of people whom they know want eggs, or even informally "take requests" from people they know might want eggs (but cannot get eggs themselves) before they go out to harvest. Such behavior should not be interpreted as a solicitation of business; rather it is merely an attempt to calibrate the level of harvest needed to satisfy demand or to estimate operational logistics in relation to planned egg distributions (number of hemlock trees to cut, number of sets to make, time to set, harvest and process, shipping costs, etc.).

Transactions of a different (non-subsistence) order, however, may develop when advertising or direct solicitation is involved. To take another hypothetical example, say Harvester C posts signs stating, "Herring Eggs for Sale —$125 a 50 pound box," in a prominent public location. This clearly goes beyond "taking requests" informally among one's Elders or relatives and could be construed as staging a commercial transaction. The sociocultural aspects of such an exchange are clearly subordinate to the instrumental economic ones. The investigators heard several stories of such direct solicitation, including one communicated in advertising signs at the Alaska Native Medical Center. However, we were not able to verify any concrete details of such bold marketing or anyone who had or would engage in it. All harvesters we talked to would only consider trade for cash with people they were at least familiar with, if not directly than through a friend or relative (cf. Schroeder and Kookesh 1990:39). Furthermore, the consensus among interviewees was that the number of such "blind selling" or solicitation cases is small (though perhaps growing) and that the quantities of eggs involved are typically not significant, as compared to the other forms of subsistence exchange. This conclusion is consistent with the broader subsistence literature on distribution and trade, which suggests commercial violations of customary and traditional exchange relations are infrequent and often censured communally (Wolfe, et al 2000). In general, as this analysis of distribution will show, people hold a general distaste and aversion for selling eggs for profit.

The Context for the Study

Prior to herring population crashes beginning in the 1940s, there was no science-based management of herring in Southeast Alaska. In fact, herring were considered superabundant if not inexhaustible. Quotas were introduced in 1940, after the first major population crash. It was too late, however, as herring had become severely overfished, and subsequently they have never recovered to their pre-1940 levels. Increasing population fluctuations, smaller (fewer mature) spawners, shorter durations of spawning and reduced extensity and intensity of spawning have characterized the fishery ever since. These changes, when combined with the fact that herring are migratory, have proven a challenge for modern fisheries managers, who rely on biomass estimation models to predict herring returns to local spawning areas. Biomass models, in combination with Maximum Sustainable Yield (MSY) guideline harvest levels for commercial fishing, have been unsuccessful in maintaining or restoring healthy herring populations throughout Southeast Alaska and British Columbia. Instead, this management regime has led to the closure of all commercial sac roe fisheries, except Sitka Sound, and an increasing concentration of herring into just a few spawning areas, with Sitka Sound being host to the largest remaining spawning population in Southeast Alaska.

To date, scientific studies have not been able to conclusively track herring ranges, seasonal movements, and level of fidelity to spawning areas (Carls et al. 2008). Correspondingly, herring life cycles are affected by many factors ranging from habitat and water quality to levels of prey availability (e.g., krill) and predation (e.g., by whales) to more global factors, such as climate change and its impacts on oceanic conditions. These factors are not accounted for in the single-species biomass models used in state management. It is hypothesized that a herring meta-population exists in Southeast Alaska into which juveniles from various areas are recruited and entrained by older herring, a theory supported by local and traditional knowledge (LTK) (Thornton, et al 2010a, 2010b; MacCall, et al 2018). As such, there are no separate genetic stocks but, rather, only "regulatory stocks" based on their shared life histories, in particular spawning environments—though geographic populations may exhibit unique chemical signatures in their otolith bones (Carls et al. 2008; Meuret-Woody et al. n.d.).

Many Natives think it is important to maintain critical spawning areas and mature spawners to support abundant spawn in extensive areas, and to be respectful of herring during their spawning process. The heavy concentration of sac roe fishing in Sitka Sound, the only viable biomass left to exploit, stands at odds with Indigenous principles of sustainable engagement of herring, and the Sitka Tribe holds that state mismanagement of the herring sac roe fisheries has jeopardized their subsistence production and distribution system. In 2018, the Tribe and allied Native organizations sought an injunction to prevent the state from opening the Spring 2019 sac roe fishery until a management review could be carried out (see Appendix 5). This conflict, especially the impacts of management of the Sitka Sound sac roe fishery on the subsistence herring egg production and distribution system, gives rise to the need for this study to document the nature of the Indigenous system.

In summary, the distribution of subsistence herring eggs harvested from Sitka Sound is prodigious, with 87 percent of the overall harvest volume given away, on average, rather than personally consumed by harvesters and their households. The distribution is also profound in its geographical extent and multiplicity of benefits, including its cultural meanings for both the givers and receivers of eggs of this "first fish" of the spring. Herring egg distribution is defined not only by the predominance of direct sharing and exchange each season, but also by its widespread range (benefitting communities across the state), social-economic complexity (of primary, secondary, and tertiary sharing, gifting, barter, and trade), and its resilience (having been carried out for thousands of years in variable social and environmental conditions). Today, cash has become a necessary component of both the production and distribution of herring eggs, yet, as this report demonstrates, it has not distorted the customary and traditional aspects of subsistence uses and forms of exchange, despite isolated reports of commercialized transactions. Instead, as this report will show, there is a high degree of consensus among the harvester and recipient populations about the nature of herring egg distributions, including their material, social, and spiritual contributions, cultural protocols, benefit flows, and meanings. As well, Natives consistently recognize the

Roby Littlefield, Kaagwaantaan Tlingit clan, assists with the blessing of the herring, a ceremony at Herring Rock, sacred property of the Kiks.ádi clan, at the advent of spawning season. (Photo: Bethany Sonsini Goodrich, Sustainable Southeast Partnership)

importance of Indigenous social organizations and ties in maintaining the complex and resilient production and distribution networks that enable widespread consumption of herring eggs. These patterns adhere even in years where the harvest may be depressed or highly localized due to poor returns, irregular spawning, or other factors.

Contemporary stressors on the harvest and distribution of herring eggs are also discussed in this report. These include decreased opportunities for subsistence uses, the "crowding out" of subsistence users by commercial fishing interests, and management decision-making. These impacts have led to less reliable herring returns, decreased access to core herring spawning areas, increased costs in carrying out herring egg harvests per unit return, and higher costs to ship herring eggs (especially via air-freight). In assessing the contemporary distribution of herring eggs, this report takes a holistic, interdisciplinary historical, cultural, and political ecological approach to herring egg distribution that recognizes herring's foundational, keystone, and bellwether role in North Pacific social-ecological systems.

Objectives of the Study

The main goal of this study is to gain a greater understanding of the nature and significance of the Sitka Sound subsistence herring egg distribution, which occurs each spring, including the direct benefits, co-benefits, and values associated with sharing and various forms of exchange. It focuses not only on patterns of material exchange of herring eggs, but also the cultural meanings and motivations behind the massive distribution of eggs that uniquely defines this subsistence fishery and the relations of production, distribution, and consumption it encompasses.

A second goal is to examine how the changes in both the status of herring spawning schools in Sitka Sound and surrounding areas, and in the regulatory regime governing commercial harvests and subsistence uses of herring eggs in the region, have impacted patterns of distribution. And, in turn, to ask how have these impacts affected the benefits, co-benefits, and meanings of herring eggs as objects of gifting, exchange, and consumption?

To achieve these aims, the following task objectives were set:

Task 1: A literature review and re-analysis of existing interview data on the traditional Sitka Sound and relevant contiguous Pacific herring (*Clupea pallasii*) fisheries was conducted to document patterns of primary harvest, use, giving, and receiving as well as communal sharing and secondary distribution through networks to various endpoints of consumption. While the Alaska Department of Fish and Game (ADF&G), Division of Subsistence data provide basic subsistence harvest and use figures going back intermittently to 1985, the data do not cover secondary or tertiary distribution, or analyze the cultural meanings of distribution.

Task 2: Based on the Task 1, a key informant interview list and set of questions were developed for Sitka Sound harvesters, the regional center of herring egg activity, but with connections to Angoon, Craig/Klawock, Hoonah, Juneau, Metlakatla, and other relevant communities. Interviews sought to document both the giving and receiving ends of distribution as well as onward distribution from receiving communities. Forty-two interviews were carried out to document local and traditional knowledge (LTK) of the key dimensions of distribution over time, including terms and quantities of exchange, uses, values, and secondary sharing. It is recognized that terms of exchange might include sharing, gifting, and trading, as well as contributions to support the herring egg production (such as labor, fuel money, etc.). While the distribution of different products, including spawn on branches, spawn on kelp, pickled herring, etc., may follow different patterns, this study emphasizes the distribution of spawn on branches.

Task 3: Working with ADF&G, Division of Subsistence, a supplemental questionnaire on distribution was created for the 2018 annual herring egg subsistence harvest survey (Appendix 2). In addition, effort was made to re-analyze existing data on sharing and distribution from previous surveys (including relevant unpublished data).

Task 4: A list of "high distributors" was compiled from initial interviews and referrals in order to conduct detailed profiles of consistent, high distributors' primary and secondary distribution patterns to understand the supply and value chains and benefit flows created by the herring egg harvest within specific social networks. A typology of distributional types (e.g. gifts, trade, etc.) and distributional entities (individuals, households, tribal organizations, corporations, and other organizations) was developed to further analyze the relationships and particular roles occupied in the distributional networks that support the herring egg economy. Benefit flows were also tracked in this context.

Task 5: Results were compiled into this report, including a critical synthesis of the literature on herring egg harvest and distribution and an analysis of the primary interview, survey, and profile data gathered as part of this study. The report includes an assessment of trends in distribution, valuation, and benefit flows within social networks, as well as recommendations for appropriate support of the herring egg social and cultural economy.

Task 6: Relevant articles and presentations from the results of this research will be prepared for dissemination to interested organizations at conferences and through scientific publication. This outreach should contribute to improved recognition, governance, and support of the herring egg subsistence economy, especially in Alaska Native communities where herring carries the status of a cultural keystone species and egg harvest and distribution is considered a foundational rite of spring (Thornton et al 2010a).

Methodology

This study employed three primary research methods. First, a targeted review of the relevant anthropological, historical, and biological literature on herring and herring fishing was carried out. This review incorporated both published and unpublished documentary records and draws partly from previous work (Thornton, et al 2010a).

Second, semi-structured, qualitative interviews were carried out following a basic set of topics and questions. PI Thornton was assisted in carrying out interviews by Forest Haven, a PhD candidate in anthropology at the University of California Irvine, originally from Metlakatla. Recruitment of interviewees was conducted through chain referrals (from Elders, tribal leaders, and known harvesters) and from a list of active herring harvesting households maintained by the STA for use in their herring egg survey with ADF&G, Division of Subsistence. We also recruited Sitka Sound herring egg harvesters from Angoon, Juneau, Kake, and other communities who travel to Sitka for the subsistence gathering, though it was not possible to interview all of them within the time frame of this study. We did not limit our sample to harvesting households, however, as distribution is as much about receiving as giving. Thus, we drew from a sample of both harvesters (28) and recipients (18, not including harvesters who also receive), the latter include Elders, homebound persons, and others who are unable to get herring eggs on their own but were known to receive them from other sources. Sampling and interviews were conducted according to ethical protocols established by local Native organizations, including the Sealaska Heritage Institute (SHI) and STA. Foundational to the ethical conduct of this study was the principle of Free, Prior, Informed Consent, which provides participants with a summary of the purpose of the study, how it will be used, and their rights to participate in the study or not, anonymously or not. Participants were asked to sign the consent form or provide oral consent, which the researchers documented. A copy of the consent form is in Appendix 3.

Typically, interviewers began by asking participants generally about their involvement in the herring fishery and the nature of their distribution and reception of herring eggs both locally and beyond, and then probed their knowledge of secondary and onward distributions of eggs to other recipients. This was important because secondary and tertiary sharing of eggs, which are a critical part of the distribution regime, had previously not been well documented in the literature or ADF&G-STA surveys. Interview topics and questions were closely tied to project objectives of documenting consultants' understandings of herring distribution patterns and their significance, and also what changes in local patterns of distribution had occurred, including changes over time and in response to social, ecological, and regulatory conditions. In total, 52 interviews were conducted with people from the communities of Sitka, Hoonah, Juneau, Kake, Angoon, and Craig, 43 of which were recorded. Nearly all of the 52 interviews were conducted individually, though an exception was the group interview conducted at the Hoonah Senior Center, during and after lunch, which proved the most expedient and comfortable way to converse. STA 3,000+ Alaska Natives of Sitka, and the Hoonah Indian Association (HIA), the federally recognized government of the 600+ Natives of Hoonah, also provided information on their herring egg distribution programs, which support the gathering, processing, and distribution of thousands of pounds of herring eggs to residents of their communities.

All interviews were conducted in English, although Tlingit was used occasionally to introduce or emphasize cultural concepts or key terms. The format of the interviews was semi-structured. Typically interviews began with a broad question, such as, "What is your relationship to the subsistence herring egg fishery?" and then moved to specific topics depending on whether the interviewee was a harvester ("To whom do you distribute herring eggs?" etc.) or a receiver ("From whom do you receive herring eggs?" etc.). The semi-structured approach allowed us not only to probe the kind of distributional relationships that are involved in the subsistence fishery but also why and how they had evolved (or devolved) over the years.

Finally, a third method used was to construct a series of individual and organizational herring egg distribution profiles which sought to go deeper into the details of how eggs are shared and why. While the ADF&G-STA survey concentrates on the basic characteristics of the distribution regime, the semi-structured interviews we carried out allowed for deeper inquiry and the ability to connect other topics and meanings to the acts of giving and receiving herring eggs, which would not be possible to capture through a simple survey or standardized questionnaire. These profiles are presented in Chapter 5. In some cases, follow-up questions were required, and several return trips to Sitka were made in order to verify information or probe topics that arose in other interviews.

Organization of the Report

The report is organized as follows: Chapter 1 provides an introduction to the phenomenon of herring egg distribution by subsistence harvesters in Sitka Sound and discusses research objectives and methods. Chapter 2 establishes the ecological, cultural, and managerial context for the herring fishery. Chapter 3 summarizes the results of previous studies of herring egg production, distribution, and exchange. Chapter 4 examines current patterns of distribution through both the survey analyses and semi-structured interviews. Chapter 5 illustrates patterns of distribution among regular herring egg harvesters and distributing organizations, such as tribes and social service agencies. Chapter 6 examines the benefits of the subsistence distribution system within a broader, integrated, cultural context, as the herring egg fishery provides much more than just a few hundred thousand pounds of eggs. Finally, Chapter 7 offers some conclusions, suggestions for further research, and recommendations concerning how to recognize, protect, and support the benefits of the herring distributional system in relation to herring populations in Sitka Sound and other communities.

CHAPTER 2: THE ECOLOGICAL, CULTURAL, AND MANAGERIAL CONTEXT OF THE HERRING FISHERY

"Herring are the key to the ocean.
They are our buffalo."

—Mike Miller, Sitka

Photo: Bethany Sonsini Goodrich, Sustainable Southeast Partnership.

Ecological Context: Herring as a Foundation, Keystone, and Bellwether Species

Herring, along with other forage fishes, occupy a "foundation" role in marine ecosystems. Herring consume tiny phytoplankton and zooplankton at the bottom of the food chain, converting them into a concentrated, high value resource—each herring brimming with omega 3 fatty acids and other high energy and nutrient sources—which in turn support a host of higher trophic-level species. Herring are, as Tlingit and Haida fishers succinctly describe them, "the feed for everything" (Harold Martin in Thornton, et al 2008a), "the most important fish in the sea" (CS), and "the key to the whole [marine eco-] system" (MM).

As the most numerous forage fish in the northern hemisphere's oceans, both Atlantic herring (*Clupea harengus*) and Pacific herring (*Clupea pallasii*) are the dominant prey for many larger species which humans similarly depend on, from halibut and tuna, to salmon and cod, to whales and seals, to sea birds and a variety of other fauna. Their importance as food (up to 60 percent of the diet of king salmon, *Oncorhynchus tshawytscha*, for example), means herring have many predators, including a most influential one: humans. Millions of tons of herring are harvested each year by fishers from all over the northern hemisphere and beyond. In this respect, herring can also be considered a cultural keystone species (discussed further below).

A sudden decline or disappearance of this foundation species can lead to the collapse or severe impoverishment of marine ecosystems, and decimated herring stocks can take decades to recover, even if suitable habitats have been restored. In Prince William Sound, Alaska, for example, where one of the largest oil spills in history took place in 1989 when the Exxon Valdez ran aground releasing more than 11 million gallons of crude oil, the herring population collapsed. Three decades later, despite significant habitat restoration efforts and careful monitoring, herring populations have yet to recover (Brown et al. 2002, EVOSTC 2011). Consequently, the marine ecosystem remains significantly less productive than in the pre-spill era. Native populations and scientists alike see the status of herring in key spawning and massing areas like Prince William Sound as bellwethers for the health and status of marine ecosystem as a whole (Thornton et al. 2010a). It is significant that, despite their importance to the marine ecosystem, herring have not actually been designated as a forage fish by the Board of Fisheries. In 2016, STA made a series of proposals to the Board to list and manage herring as a forage fish and also ran an online petition campaign in support of the proposal, which garnered significant support. The Board of Fisheries chose not to adopt the proposal, however, which would have necessitated taking a more conservative approach to managing herring (see Board of Fisheries proposals 209, 210, 211; STA 2016).

Cultural Keystone Species, Biocultural Diversity, and Ecosystem Services

Herring can be classified as a cultural keystone species (Garibaldi and Turner 2004) among many coastal peoples of the north Atlantic and Pacific oceans (Thornton 2015). Cultural keystone species are characterized by 1) intensity, cultivation, and multiplicity of use, 2) rich linguistic and 3) cultural associations (e.g., subsistence dependence, mythology), 4) persistence in memory and use despite cultural change (e.g., through traditional environmental knowledge), 5) unique and irreplaceable role in socioecological systems (e.g., as a foundation and bellwether species), and 6) value in providing opportunities for resource acquisition beyond the home territory (e.g., through trade or exchange). In northwestern North

America, the archaeological record of faunal remains shows that herring were fished for more than 10,000 years and were routinely taken by at least 4,000 BP (Thornton et al. 2010a; Moss et al. 2016a).

Because herring are migratory, people took note of their annual cycles of movement. Herring schools typically move to deeper waters offshore during the winter, returning in late winter or early spring to spawn in favored intertidal areas along the Southeast Alaska's rocky coastline. Their return was celebrated, and the spawning of billions of eggs marked the beginning of the productive season for fishing, gathering, and other food collecting in many coastal Indigenous communities. Salmon, halibut, and seals followed the herring to their spawning grounds, increasing the bounty and diversity of species in coastal ecosystems. Indeed, the re-appearance of herring in spring brought the coastal seascape alive again, and the presence of adult and juvenile herring and their spawn and developing larvae kept it vital through the height of the summer salmon runs.

Thus, herring were esteemed not only for their food value—as eggs, meat, oil, and mash—but also for the ecosystem services they provided. Herring attracted and supported other coveted species, such as halibut, salmon, sea mammals, seabirds, and marine invertebrates. As Tlingit Elder Harold Martin (in Thornton, et al 2010a:87) explains: "[Herring are] important to everything...they're the feed for everything...[W]e didn't like the idea of commercial fishermen coming in and taking them on a large scale because they're very important to our salmon and especially king salmon, you know. And they feed our seals and stuff like that—things that we're depending on." Elders throughout Southeast Alaska and related areas bemoaned the decline of regional herring populations which has become a concern over the past three quarters of a century due to industrial fishing, beginning with the reduction era, and has not seen significant recovery to the present day. Due to their deep knowledge of, dependence upon, and intimate relations with herring, Southeast Alaska Natives have long been concerned about the potential for damage to the resource through overfishing, spawning disturbance, and habitat degradation.

Part of this concerns is motivated by a clear understanding of the importance of herring to biological diversity and a healthy marine ecosystem, which, in turn, supports cultural wellbeing and diversity. Indeed, on the Pacific Coast of North America and, to a lesser extent, Northeast

The waters of Sitka Sound historically turned white with herring milt during spawning. Today this spectacle is more limited, often to just a few islands and bays. This contraction of spawning extensity concerns herring egg harvesters. (Photo: Bethany Sonsini Goodrich, Sustainable Southeast Partnership)

Asia, the presence of herring correlates with high biological and linguistic diversity as well as high levels of trade and extensive resource distribution (Suttles 1990). Some of the densest and most diverse populations of hunter-gatherer-fisher groups are associated with areas rich in herring massing and spawning. On the Northwest Coast of North America between Southeast Alaska and Oregon, more than 50 distinct language groups have evolved, nearly all of which had relationships with herring and relied on the bountiful marine ecosystem herring supported.

In terms of distinct ethno-linguistic groups, this area of the Northwest Coast is one of the most culturally diverse regions on the American continent. If the herring-rich areas of coastal California are included, it is the richest area in terms of ethno-linguistic diversity. Coastal Siberia contains fewer Indigenous languages (10) but still a comparatively high level of ethno-linguistic diversity. Most languages in both areas are now considered endangered, rendering both coasts of the North Pacific "hotspots" for language (Davidson and Harrison 2006) and biocultural diversity conservation. While the reasons for this decline of linguistic and cultural diversity are multivariate and complex, and also linked to multiple forms of colonization, it is important to keep in mind the significant correlation between the wellbeing of marine ecosystems and the biological and cultural diversity they support.

There is further evidence from documentary, archaeological, and oral historical sources that societies distributed themselves strategically in relation to areas rich in biodiversity associated with Pacific herring. Groups situated near spawning areas developed techniques for managing them by means of cultivating supply and regulating demand. Techniques to regulate the supply of herring included preservation and storage, trade and exchange, habitat conservation, habitat cultivation, return of selected egg deposits, transplantation, and efficiency innovations. On the demand side, key regulatory controls included territoriality, mobility, prescriptions and prohibitions on herring interactions, substrate choice, and sabotage or sanction of greedy harvesters or hoarders (Thornton 2015). While groups dwelling in areas with abundant herring came to specialize on them, neighboring peoples developed trade or participatory exchange relations to gain access to the fisheries, or in some cases successfully transplanted herring to their own areas.

Given their deep historical and wide geographical status as a foundation, bellwether, and keystone species for bioculturally diverse marine ecosystems, their value as an immediate, storable, and tradable foodstuff, and the intimacy with which local peoples are familiar with ecology, there is considerable ecological knowledge, practical skills, social networks, and cultural capital associated with herring spawning and harvesting. This report focuses on the cultural values associated with the distribution and exchange of herring, which have been built up and handed down from generation to generation.

While many aspects of cultural life and fisheries in the Pacific Northwest have changed dramatically since contact, the nature and culture of herring distribution has persisted in form, if not abundance. Sitka subsistence herring egg distributor Mike Miller observes:

> I've seen a lot of changes over the years, both in the fish and the fishery, and the only thing that's been kind of a constant is the traditional harvest of eggs and the sharing of Sitka herring eggs around the state. Interestingly enough, I share eggs with people as far north as Barrow and places that don't have herring and don't have trees, but they like herring eggs...We have quite an extensive trade network because they'll send whale meat from Barrow to me and I'll send them herring eggs. And some places, they have different kind of fish than we do and different kind of salmon. So I'll send them herring eggs and they send salmon...About 20 years ago now I was trading ...with some people in Washington State and there was a friend of mine and his mom—at the time she was about 100 [years old]...and she remembered that when she was a little girl they used to trade for herring eggs. And this was eastern Washington; it's not on the coast. And ...supposedly they would bring canoe loads of herring eggs.

This statement attests to the cultural keystone status of herring throughout most of the Northwest Coast culture area and adjacent plateau and interior communities with which they traded, as well as the persistence of these links.

A fishery like Sitka Sound, with an abundance of herring to support a rich marine ecosystem and wide distributional network, is recognized and valued for the bounty it provides. Nowadays, these benefits are often described as "ecosystems services," following the Millennium Ecosystem Assessment (MEA 2005). Ecosystem services include provisioning services, such as the production of food; regulating services, such as the control of climate and disease; supporting services, such as nutrient cycling; and cultural services (typically limited to intangibles but including values associated with exchanging food), such as spiritual and heritage benefits. Herring egg production for trade and distribution can be characterized as a provisioning service and a cultural ecosystem service. The Millennium Ecosystem Assessment demonstrates how ecosystems sustain human wellbeing by providing these four critical ecosystem services. Proponents of biocultural diversity (cf. Maffi 2005; Maffi and Woodley 2010), emphasizing the vital links and corresponding losses between biological diversity (number of endemic species) and cultural diversity (number of Indigenous languages), argue for recognition of the foundational relationships between the wellbeing of Indigenous peoples and the health of the diverse ecosystems they inhabit and the ecosystem services these cultural landscapes provide. Sitka Sound is a classic example of a rich cultural landscape that correlates with its rich and diverse ecosystem services, a critical foundation for which is herring.

Historical ecologists tend to view the relationships between biological and cultural diversity not only as correlative but as co-evolutionary (cf. Balée 1998). In other words, diverse aboriginal societies and the landscapes they inhabit have co-produced each other over long periods of time through the dynamics of interaction, use, and adaptation. This means that local peoples not only recognized and valued ecosystem services, but also helped to cultivate key species through "services to ecosystems" (see Comberti et al. 2015). Unfortunately, the role of aboriginal peoples in producing and maintaining the world's varied ecosystems and their biological diversity has not been recognized by mainstream development or conservation organizations, which have generally pursued models of intensive exploitation of marketable resources in the form of development, or attempts at "fortress conservation" in the form of strict preservation (Brockington 2002; Brockington et al. 2008; Dowie 2009). The Alaska constitution and statehood movement was among the first to call for a more sustainable model of fisheries development for the long-term benefit of local peoples. Significantly, the constitution (Article 8.3) also stresses "the utilization, development, and conservation of all natural resources belonging to the State, including lands and waters, for the *maximum benefit of its people*" (emphasis added).

Today, Pacific herring fisheries are considered to be well managed by government and industry, and even independently certified as "sustainable" by the Marine Stewardship Council (MSC 2017). Yet, the Sitka Tribe and other observers aver that herring management in Southeast Alaska is not providing adequate benefit flows to meet Native subsistence needs as well as the needs of other species they depend on (via ecosystem services). Thus, the question of long-term health and so-called "shifting baselines" in herring populations—from a state of superabundance at historical contact, to one of scarcity in the present day—remains a critical one (Pauly 1995; Roberts 2007; Thornton and Hebert 2015).

The State Managerial Context

The Alaska Department of Fish and Game (ADF&G), established with statehood in 1959, manages herring fishing in Alaska with oversight and guidance from the Alaska Board of Fisheries, a seven-member citizen board established in 1975, with members appointed by the Governor and approved by the Legislature. Officially, ADF&G recognizes the social, cultural, and economic importance of subsistence herring egg fisheries in the Gulf of Alaska and Bering Sea:

> Subsistence fisheries for herring in Alaska predate recorded history. The spring harvest of herring eggs on hemlock boughs or kelp has always been an important subsistence resource in coastal communities throughout Alaska. Traditional dried herring remains a major staple of the diet in Bering Sea villages near Nelson Island. Pacific herring are also harvested by subsistence users for

> consumption as fresh fish and for bait (http://www.adfg.alaska.gov/index.cfm?adfg=herring.uses).

However, management to date has been driven by commercial fishing interests. The timing, spatial prescriptions, and other regulations surrounding herring fishing are largely determined by commercial needs since the expanded industrialization of the fishery in the early 20th century. With the State of Alaska's takeover of commercial fisheries in the 1960s, management of the fisheries has evolved according to the MSY model, as per the State's constitutional prescriptions.

The MSY method involves first estimating the "biomass" of herring in an area and then determining the "harvestable surplus" of the biomass that can be taken without threatening the sustainability of future herring returns. ADF&G has developed a methodology for estimating the herring biomass and assessing what percentage of it can be harvested, as follows:

> Historically, ADF&G has conducted two direct observational methods for estimating herring biomass in Southeast Alaska: (1) egg deposition dive surveys, and (2) vessel hydroacoustic surveys. In cases where egg deposition surveys are used, the biomass estimate is based on data from mature herring that spawned that season. Acoustic surveys also have been used to estimate biomass, but this method is thought to be less reliable than egg deposition estimation.
>
> Herring stock assessment activities begin in the spring with aerial and foot surveys conducted by department staff. Herring spawning activity is documented during these surveys and used to determine, among other things, where to distribute SCUBA dive transects. The dive transects are used to derive egg deposition estimates which are used along with many other biological parameters, including age and growth, to estimate spawning biomass. Age and growth information are obtained by samples collected from test fishing, commercial harvests, mid-water trawling, and sampling on the spawning grounds by the department.

> Forecasts for herring in Southeast Alaska are currently computed using either an age-structured assessment (ASA) or a biomass accounting method, where the observed spawning biomass and age composition from one year is modified by estimates of growth and mortality to produce a forecast for the subsequent year's biomass (ADF&G 2019: http://www.adfg.alaska.gov/index.cfm?adfg=herring.research).

This assessment model is controversial, however, because it has not produced consistent sustainable surpluses for harvest. In fact, most of the commercial herring management areas have been closed due to poor herring returns (Thornton and Hebert 2015). Furthermore, Sitka Sound, regarded as the largest and healthiest of the spawning herring populations, is viewed by most locals as existing in a depleted state when compared to historical herring spawning schools which were larger in total size, contained larger individual spawner fish, and spawned for longer durations across a much wider range of shoreline than in the present day. Many long-term observers perceive a fishery that is being managed in a way that maintains a "shifted baseline" (Pauly 1995).

What is meant by shifting baselines? In a nutshell it is the phenomenon wherein historical consciousness of the size and productivity of a fishery generations ago is lost to memory, and participants (be they fishers or managers) come to accept the new, diminished state of the population as "normal." In the absence of good scientific records, and the advent of new technologies (which continue to find fish despite their declining numbers), human perceptions can indeed shift, and managers can take the "shifted baselines" as management targets. When combined with MSY harvest policies, the results can leave little margin for error, and overfishing can continue to undermine already vulnerable fish populations. This is especially important among forage fish fisheries, which continue to be managed according to reductionist, single-species models of productivity, or biomass, and MSY exploitation rates. According to Todd (2017:48-49), even commercial sac roe fishers have concerns about the viability of the fishery under this model.

This MSY model for commercial herring production can be contrasted with Indigenous Tlingit and Haida cultural models of cultivated abundance and ecosystem stewardship, which take a conservative view towards the harvesting of herring on the verge of spawning, and instead favors helping them to spawn successfully by laying hemlock branches in the water for spawning substrate, transplanting eggs, and restricting noise and other distressing interactions during spawning (Thornton 2015, and below). This model cultivates abundance; however, it was originally not "legible" (Scott 1998) to the state, and while awareness of these cultural models has increased, they remain untranslatable within the narrow biomass productivity frame.

In contrast to the biomass framing of herring conservation by fisheries biologists, STA maintains that a healthy spawning population is characterized by its broad distribution in time (duration of spawning) and space (areas associated with spawning over generations). According to STA, subsistence harvest of herring eggs (which the State has an obligation to maintain as a priority over other consumptive uses) requires "a span of no less than 3 days of spawn in one particular area" (STA 2009). In fact, several weeks' duration of herring spawning was the norm in the past (Thornton et al. 2010a). The Tribe further argues, based on local and traditional knowledge, that if a commercial sac roe fishery is initiated prior to spawning, large schools of spawning herring are disturbed, adversely affecting subsistence herring egg harvests.

The spatial boundaries of the commercial fishery are also integral factors to a healthy herring population. In 2008 more than 10,000 tons herring were taken in a single commercial opening in a confined part of Sitka Sound. A concentrated harvest of spawning herring of such magnitude, argues the Tribe, may "irreversibly" disrupt a herring stock by removing "the repeat spawners that would have had knowledge of migration routes and selecting spawning habitats" (STA 2007). Targeting spawning stocks within a concentrated area of Sitka Sound, the Tribe suggests, is having progressively adverse effects not only on herring schools' abilities to carry out spawning throughout their historical range, but also on Tribe members' abilities to meet their subsistence needs (up to 300,000 pounds of eggs per year; see Appendix 1; Holen et al. 2011; STA 2009).

Prior to 2002, there was no formal consultation with the Tribe regarding their local and traditional knowledge of Sitka Sound herring, despite numerous complaints raised by the Tribe and its citizens about the management of the sac roe fishery (Thornton, et al 2010). In 2002, in response to conflicts brewing between subsistence users and ADF&G fisheries management, the Board of Fisheries requested that ADF&G set up an improved consultation framework for addressing subsistence herring egg gathering concerns, and the ADF&G commissioner entered into a memorandum of agreement (MOA) with STA to do so. The MOA did improve communication but did not address the majority of STA's concerns, and ADF&G unilaterally terminated the agreement in 2010. Subsistence herring egg harvesters continued to testify to the damages that the sac roe fishery was wreaking on their subsistence uses, resulting in frequent failures to meet their gathering needs (Craig in Thornton, et al 2010a: Appendix D).

The State's herring management plan for Southeast Alaska (5 AAC 27.190) includes minimal mechanisms for mitigating impacts on subsistence egg harvests, such as limiting harvests in spawning areas with the highest frequency of use and dispersing fishing effort in time and space (ADF&G 2009b). Only in 2012, however, did the Board of Fisheries pass a regulation that had been pushed for years by STA, mandating that the core area for herring spawning and subsistence egg harvest be closed to commercial fishing. Nonetheless, the closed area is small, and commercial sac roe seiners still have de facto priority in capturing the surrounding herring schools which are found and bound (by area markers) for them by State fisheries managers (Thornton and Kitka 2015). This framing reinforces a perception among many subsistence users that, contrary to the intent of subsistence laws, herring fishery regulations clearly favor commercial sac roe harvesters uses over subsistence egg gatherers' needs.

Commercial fishing of herring roe in Alaska is not community based, but rather a state managed, limited-entry system, which regulates privatized, marketable permits for the sac roe (primarily Sitka Sound) and spawn on kelp (primarily Craig) fisheries. Commercial fishers compete head-to-head with subsistence fishers in both time and space on the only significant remaining spawning aggregations in the region. While ADF&G has acknowledged "that large numbers of vessels and increased

Herring sac roe seine fishery, Sitka Sound. This fishery takes more than 99 percent of the herring harvested in Sitka. (Photo: Bethany Sonsini Goodrich, Sustainable Southeast Partnership)

activity on the spawning grounds may disrupt herring activity and cause them to move and spawn in other areas," (STA 2009:10) in concurrence with STA and local and traditional knowledge, the tight spatial framing of the fishery on the concentrated spawning stocks in Sitka Sound makes this conflict inevitable. Further, the temporal imperative to take commercial roe at the threshold of spawning means that subsistence harvesters are left with access to only what remains of this increasingly circumscribed local spawning population after the commercial capture. As a practical matter, then, commercial fishers have priority access to the herring, save for the small closed area designated in 2012 and enlarged slightly in 2018.

As a result of this growing competition and increasing subordination of subsistence fishing to commercial fishing, Sitka Natives have not been able to meet their subsistence needs for eight of the 16 years since 2002 when subsistence harvest surveying has been carried out between

STA and ADF&G, including seven of the past 10 years (2007-2016; Appendix 1). It is too early to tell if the situation will improve with the successful implementation of the subsistence-only (commercial no-take) areas defined in 2012 (and expanded slightly in 2018) that encompass some of the traditional core subsistence gathering areas in front of town. More effective measures will likely be necessary to reduce the percentage of the herring biomass commercially fished (perhaps to a precautionary 10 percent) and increase protected areas (i.e., subsistence-only fishing areas) in order to reverse the current stress on herring stocks and subsistence egg gatherers.

The Cultural Context: Herring as Keystone

Pacific herring have long been prized among Southeast Alaska Natives for their massive return in spring to break the winter scarcity of fresh food and herald the return of the prodigious runs of salmon to their natal streams. "Herring are our Easter eggs," analogized one Tlingit Elder (JL), emphasizing how they literally (rather than the more symbolic "bunny eggs") mark the spring rebirth and renewal of life. Herring provided fresh meat, oil, and eggs, which were consumed fresh or put up for storage or trade. As much as two to three months before laying their eggs in late March or early April, herring moved into spawning areas in large schools. Concentrated thus in coastal bays and inlets, they could be raked, jigged, netted, or trapped to be eaten fresh, smoked, or rendered into oil. In autumn, when salmon runs tapered off, herring could still be harvested for meat and oil in areas where they massed, such as Angoon, in Southeast Alaska (Figure 1.2; de Laguna 1960; Emmons and de Laguna 1991).

The origins of herring fishing are deeply rooted in legend. A Haida version has the trickster-demiurge Raven discovering the best technology for acquiring herring eggs as part of his epic and insatiable quest for food (Swanton 1905a:135). He used a kelp colloquially known as "Raven's moustache" as his substrate to capture the eggs. From Seagull he learned how to capture live herring when they returned to spawn by tricking the gull into demonstrating to him how to capture herring and then stealing his catch (Swanton 1909:13-14).

According to Elder Herman Kitka Sr., Tlingits trace their name for herring (*yaaw*) to seagulls (specifically the glaucous-winged gull, *Larus glaucescens*), which audibly change their cries to a satiated "*yaaw*" once herring arrive to a spawn. At the Tlingit community of Sitka, herring eggs were said to have been first collected on the hair of a young woman as she lay sleeping upon Herring Rock (*Yaaw Teiyí*; see Thornton 2008). "She'd go down there when the tide would be coming up," Elder John Nielson recounts, "She had real long hair. About that time the herring would start coming in. They'd mill around her hair and they'd start spawning in her hair [perhaps taking it for hair kelp]. Then they found ... there's some food value in that. So they start gathering [herring eggs] ...on the hair kelp. And then we start setting branches too" (Thornton et al. 2010a). Herring Rock was taken as a crest by the Kiks.ádi clan, and names were given to commemorate the woman and the event (*Yaaw Sháa*, "Herring Woman," or "Herring Maiden," sometimes collectively applied as *Yaaw Sháawu*, "Herring Women," or *Kax̱átjaa Sháa,* "Flipping Girls"), names which continue to be passed on today. From these origins, Southeast Natives further evolved cultivation techniques for enhancing herring productivity including the laying of hemlock branches and kelp as substrate for herring spawn (Thornton 2015).

Tlingit and Haida clan groups regulated the access to herring fishing sites within their territories. Given their historical superabundance, however, restrictions on access to herring fisheries were few, except where human activity might disturb spawning (see Thornton et al. 2010a). Thousands of non-locals regularly came to the major herring fisheries such as Sitka to participate in the egg harvest. These events were celebrations in which a range of social and economic exchanges were carried out.

Early accounts of Southeast Alaska Native herring fisheries come from Russian colonizers of Tlingit territory at Sitka, or New Archangel, the most productive herring fishery in Southeast Alaska. Lütke (1987 [1835]), who was at Sitka in 1827, reported that during the spring herring season, the Natives used to gather near the fortress of New Archangel [Sitka] up to a thousand strong, and an equal number were on the nearby islands. In summer, there were often five to six hundred Natives. At first none was permitted to camp near the fort, but Governor Mouraviev (1821-27) finally changed this policy, thinking it more advantageous

Invitation to a Herring Festival and Koo.éex' [K̲u.éex']("Invite Party") at Sitka during the 2019 herring spawning season, issued by The Herring Rock Water Protectors, a group of Sitka Tlingit and other Natives and non-Natives, modeled after the Standing Rock Movement (against the Dakota Access Pipeline), which seeks protection of herring and subsistence uses of herring. The emblem of the group depicts the Kiks.ádi sacred site (*at.óow*) of Herring Rock (*Yaaw Teiyí*) surrounded by the "Flipping Girls" (*Kax̲átjaa Sháa*), female herring with whom the famous Kiks.ádi woman engaged (see https://vimeo.com/253758747 for more information).

to have the Natives and their families under the guns of the fort. From that time on, according to Lütke, the Tlingit became more peaceful. Tikhmenev (1978 [1861-63]:368, 422), reporting on conditions circa 1860, mentioned that the herring fishery at New Archangel was very good, and that in February and March, when most herring were taken, the water became milky from the spawn and milt, and that it was easy to catch the fish with a rake. Herring were also caught at Redoubt Bay in December-January (Grinev 2005) and in the autumn (Emmons and de Laguna 1991:118-119), especially at Angoon, where they were rendered into oil and mash. Early accounts of herring fishing in Sitka are further summarized in Schroeder and Kookesh (1990) and other sources. However, these sources say little about the nature of distribution.

Early Non-Native Accounts of Herring Fishing Techniques and Distribution

In Sitka, egg drying was done in special "herring trees" that featured branches extending parallel to the ground (trained to maintain their shape) and exposure to breezes. The dried eggs could be pressed into cakes but were more commonly packed away in containers for future use. The eggs could be preserved for months and reconstituted by soaking them in saltwater and then blanching them in steaming water.

A detailed account of herring production in the late nineteenth and early twentieth centuries is provided by George Emmons (Emmons and de Laguna 1991:117-119), who spent considerable time in Sitka during that period. Although well aware of Sitka's famous herring spawn, Emmons also identifies other important herring areas, such as Angoon's Kootznoowoo Inlet.

> The Alaska herring, yow, yar [*yaaw*], though a small fish, is extremely rich in oil and was greatly valued by the natives on this account, as well as for its spawn. Its distribution is limited, and it cannot be depended on from year to year except in a few localities, of which the chain of lagoons penetrating the western shore of Admiralty Island at Kootznahoo Inlet was the most favored. Vancouver's surveying party [Vancouver 1801, 5:422] mentioned the great abundance of this fish found here in July 1794, and this abundance has continued to this day. In April, vast schools enter the bays and inlets among the islands to spawn, and they deposit their eggs, khouk [*gáax'w*], in countless millions on every growth of sea plant and on the rocks in comparatively shallow water. Taking advantage of this, the natives cut hemlock boughs which they placed on the shore at low water, and weighted down with boulders. Or, still better, the branches were anchored with a rock tied to the butt ends, so that in standing up they would present more surface for the deposit of the eggs which adhered in great clusters. A broad-leafed

seaweed, dow [*daaw*, macrocystis kelp], "feather" and a long black skeleton moss [or sea grass], nh [*né*, "hair in the water"] seemed equally desirable to the fish to spawn upon, and for the natives to gather when so covered. These were fished for from the canoe with a long spruce pole with a small crosspiece seized at the end, called "sea moss catcher," nh heetar [*né xit'aa*, "sea-hair broom"]. This was entangled in the moss, with a sudden twist the root was detached, and it was brought to the surface and deposited in the canoe. ...

After spawning, the herring disappear into deep water, and about September, following their food, a tiny crustacean, they again school in myriads in the inland waters, but not necessarily on their spawning grounds from year to year, but always about Kootznahoo Roads. For this reason the resident natives of Angoon are the herring fishermen of the coast. At this season and extending through the fall, the fish are very fat and are valued most highly as oil producers, although also enjoyed as fresh food. They were sun dried in limited quantities.

Fishing was done in small canoes. The steersman in the stern paddles to where he saw the fish jump or the seabirds gather, since birds, larger fish, and men were all equally alive to the abundance that nature offered them. The bowman used a rake, heet lar [*xídlaa* or *xítlaa*]. This was a spruce staff about twelve feet long, oval at the handle, and flattened at the other end to a thickness of two or more inches. Along one edge [of the flattened end], for a distance of three feet, were inserted sharpened nails, an inch or so apart. Iron has taken the place of older bone teeth. This implement was driven down into the water with a long paddle stroke and brought up with its teeth up. With a quick turn and jerk over the canoe, the impaled fish were shaken off.

His detailed account is noteworthy in that it highlights Sitka as a dominant locale for egg production and distribution, and Angoon for oil production from fall herring fishing. Emmons' account also lists the dominant subsistence methods and means of processing. He does not discuss distribution in detail, however.

Additional details of herring processing and consumption are provided by Emmons:

> Eulachon and herring were the only other fish [besides salmon and halibut] that were cured for winter use, but only in limited quantities, for their importance as oil producers was paramount, and only the surplus was preserved for winter food. Both were also eaten fresh. [It will be remembered that the herring which were caught in the fall were fat and so were used for their oil.] This trying out of herring oil, both for home consumption and for trade, was carried on only by the Hootzah-tar Qwaan [Xootsnoowú K̲wáan] who built very substantial houses along the shores of the lagoons of Kootznahoo Inlet where they spent the fall. The fish were tried out in canoes with heated stones. ... But since this was more of a commercial industry, several special implements were used. Thus, a long-handled wooden straining spoon with slits and holes in the bow was used to remove the mass of fish from the canoe. This was emptied into a long, open-weave, spruce root bag. When half filled, the bag was folded over and laid on a slat frame over a large box half filled with water. On top of the bag, a heavy slab of wood, two feet square and two inches thick, with a handle at one end or corner, was placed and pressed down. The oil was skimmed off the water in the box with a wooden ladle, often finely carved, and was put in boxes about one-foot square and eighteen inches high, as yellow grease. It was eaten with all dried foods: fish spawn, berries (except for the soapberry), potatoes, and seaweed—in fact with every kind of food, and if it were rancid, this added rather than detracted from its quality. It was largely used

in trade and had a fixed value. With the establishment of the Killisnoo Oil Company [probably the Alaska Oil and Guano Company that had acquired the old whaling station at Killisnoo in January 1887 and was processing herring oil], the natives gave up its manufacture, since they could procure a better product [from the company] with a minimum cost of time and labor.

A limited amount of herring was cured for home consumption. The woman preparing the fish held it in her left hand just below the head, belly up and head to the left, and with a knife cut the fish from the gills along the stomach. With the thumb of the right hand she pushed out the entrails and this was all the cleaning done. The fish were strung on a small alder stick or line, passed through one gill and the mouth, alternate fish being strung through the right and through the left gill, so that the fish stood out from each other when the string was placed over a rod, and thus had a better opportunity to dry. After being sun dried for a day or two, they were hung under and about the smokehole of the house, and so cured. Or, they might be strung together with lines, fastened alternatively around the head and small of the tail and crossing between each fish; or the fish might be opened lengthwise and strung through the head. Like all other dried fish, they were dipped in oil when eaten.

Herring eggs [procured in the spring] were considered a luxury rather than a regular article of the diet. When eaten fresh, they were dropped into boiling water for an instant. [To cure the eggs], the moss, seaweed, or boughs to which the eggs adhered were hung on the fish racks, or on lines, trees, and bushes, and sun dried. They were temporarily put in bags or boxes, as the [spring] season when the spawn is very short, and the cleaning process was not attempted until later. ... To prepare the eggs for winter use, they were put in a box of water, together with the material to which they were attached, and when the eggs swelled, they were stripped

> from the broad leaves of the seaweed or from the moss. Spawn on sea moss was often packed away as gathered, since the moss does not decay and is tasteless. The eggs on hemlock branches had to be stripped very soon, before becoming impregnated by the pungency of the needles. This was done by breaking the twigs, rubbing them together, and dropping them in water, where the eggs separated and sank, while the needles and woody particles floated to the surface and were easily removed. After the eggs had been separated and cleaned, they were sun dried on boards, canoe covers, or so forth, and then packed away in boxes for winter.
>
> Cured eggs were allowed to boil in a little water for a few minutes, and were generally eaten with oil or grease. Herring spawn constituted an article of trade, and in 1882 were worth $3 for the amount contained in a hard-tack box (Emmons and de Laguna 1991:145-148).

Emmons considered herring eggs to be a "luxury," in part because they were available for such a limited period (thus rare) but also because so many were gifted or acquired through trade for other "luxury goods." Yet, herring eggs did not need to be given away due to a short shelf life, as they could be preserved—through drying, traditionally, and now by freezing—for use long beyond the spring harvest time. As such, herring bolstered Native food availability throughout the year, especially when other fish and resources were less accessible. In Sitka, herring eggs were (and remain) a staple and iconic food at ceremonial occasions, such as *ku.éex'*, where large quantities are consumed.

An early twentieth century account of the 1914 subsistence fishery was made by a federal official named Inspector Walker, who filed a report on "Waste in the Herring Fishery" (United States 1916). It shows the extent of gathering and distribution of herring eggs during this important transition era.

> At Sitka, the conditions were ...found to be quite similar to those at Craig [where in the vicinity of Fish Egg

Island more than 500 natives were found to be harvesting eggs from Klawock, Shakan, Hydaburge, Kake, Killisnoo (Angoon), and "many other places] except that the spawning grounds are in small isolated coves in the many islands, rocks, and reefs of the region...The period of spawning in 1914 in the vicinity of Sitka lasted almost three weeks. A large portion of the native population of Hoonah, Killisnoo [Angoon], Kake, and other villages, was busily engaged there in collecting the eggs and shipping or drying them.

The eggs are shipped in boxes of the size indicated above, or in burlap sacks which hold about the same quantity [~50 pounds]. On three weekly trips of the steamer *Georgia*, leaving Sitka on the dates given, the following numbers of packages of eggs were shipped to the places indicated:

Date	Packages	Destination	Date	Packages (50 lb)	Destination
1914			1914		
April 1	1	Chatham	April 6	14	Douglas
" "	12	Killisnoo	April 12	1	Chatham
" "	2	Tenakee	" "	19	Hoonah
" "	12	Hoonah	" "	3	Haines and Klukwan
" "	2	Funter	" "	19	Juneau
" "	7	Douglas			
" "	20	Juneau			
April 6	2	Hoonah	**Total**	**114**	

This makes a total of 114 boxes and sacks shipped on the three trips of the one steamer. [The Inspector did not specify prices for boxes distributed from Sitka, which may have been gifts, but reported that from Craig "A box filled with fresh eggs on seaweed or hemlock, weighing about 50 pounds, sells for about $2," thus showing

> the longevity (of over a century) of what is now called customary trade for cash]. The above figures show an incalculable number of eggs...but far greater numbers were carried away in the boats of the natives who had come from other villages to secure them. Of these there are not figures available, nor is there any way of estimating the amounts kept in their village and vicinity for future use or the amounts consumed fresh, but on every hand were to be seen both old and young natives devouring the eggs fresh and cooked. By the natives they are considered a great delicacy, but to the average white person they are wholly insipid. At the Sheldon Jackson school for natives at Sitka the demands for these eggs became so insistent that the management was finally prevailed upon to serve one or more meals of them to satisfy the children.
>
> Although the operations of the natives destroy great quantities of eggs, their destruction is insignificant in comparison with the natural enemies and the seiners...
>
> The white man's most direct and needless destruction of the herring is the seining of them on their spawning grounds and vicinity when spawning or about to spawn....
>
> The taking of herring or the disturbing of their spawn should be prohibited in southeastern Alaska each year from March first to May first. It is during this period that all spawning occurs....

The Inspector's account suggests that eggs were not merely a luxury but a critical first fish of the season. His observations on the duration of herring spawning and egg gathering accord with those of Herman Kitka Sr., a Kaagwaantaan Tlingit Elder born in 1914, who also saw the changes that came with the commercial herring reduction and sac roe eras:

> Before the commercial herring sac roe fishery, the Sitka Tlingits engaged in herring roe fishing from small rowboats. The herring caught in small rowboats were used to make herring oil. Herring roe was taken and dried in the trees among the islands around Sitka. All the small

> islands have Tlingit names and were used together with all other families in the Sitka area.
>
> Herring fishing [egg gathering] was the last week of March and the first two weeks in April, mostly under our uncles' direction for each family group. (Kitka 1998:48)

Both the Inspector and Mr. Kitka attest to the adverse impacts of the herring seine fleets on the herring egg spawning, and to the three-week duration of spawning in Sitka prior to heavy industrial fishing in the area. Mr. Kitka (here and more explicitly in other conversations with the author) reports that most able families in Sitka participated in the egg harvest, under the direction of maternal uncles, in addition to visitors from surrounding communities such as Hoonah, Angoon, Juneau, and Kake, suggesting the total number of participants during his youth was consistent with the estimates of 2,000 Natives provided by nineteenth century observers like Murawief (in Krause 1956:39), discussed in the next chapter.

As noted above in the section on ecology, Natives based their opposition to heavy seining of herring in their areas—be it for bait, reduction, or sac roe—upon the recognition of the critical foundation role of herring as ecosystem service providers, not only for humans but also as feed for other species upon which humans depend. For example, at Angoon, it was understood that spring herring returns would attract king salmon, which could be caught in the bays and inlets around Angoon in springtime well before their migration into freshwater streams for spawning (George and Bosworth 1988:53). Similarly, Klawock Tlingits told Salisbury (1962:151), succinctly, "Herring come- whale come- sea lion- seal- king salmon everything eat herring, come-big time!" Or, more recently, as Elder Harold Martin (originally from Kake) put it: "[Herring are] important to everything...they're very important to our salmon and especially king salmon, you know. And they feed our seals and stuff like that—things that we're depending on" (in Thornton, et al 2010a:78). Thus, Tlingits recognize their mutual dependency on herring with other species, such as king salmon and sea mammals.

Variations on traditional harvesting and processing techniques are further described by Lisianski (1814:239) and others (see Schroeder and

Kookesh 1990) for Sitka; George and Bosworth (1988) for Angoon; de Laguna for Angoon and Yakutat (1960, 1972); Newton and Moss (2005) for Angoon, Kake, and Hoonah; Thornton (1997) for Juneau; Ellanna and Sherrod (1986) and Salisbury (1962:151-162) for Klawock; Victor-Howe (2008) for Hydaburg; Oberg (1973:69) for Klukwan and Haines; Firman and Bosworth (1990) for Kake; and Cohen (1989) for Wrangell. These studies reveal important local and regional differences in Native production, such as the Hydaburg-Craig-Klawock preference for roe on kelp, the Sitka preference for roe on branches, and the association of Angoon with autumn herring oil production. Modern studies, such as those by the ADF&G, Division of Subsistence, also highlight changes to the herring schools and subsistence patterns, such as the advent of preserving eggs through freezing and of shipping eggs in large quantities on private vessels or via transport companies, such as Alaska Airlines (Schroeder and Kookesh 1990).

Industrial Exploitation of Herring

Intensive exploitation of herring did not commence until the American era of Alaska's colonization, specifically in 1882 with the opening of the first reduction plant at Killisnoo, near Angoon. As in Japan, the industry was stimulated by growing demand for oil in various commercial products and especially for agricultural feed and fertilizer. Seine boats up to 100 feet in length worked the anchoveta, sardine, and herring fisheries year-round, netting schools of the small oily fish from Peru and Mexico on up to California, Oregon, and Washington, British Columbia, and finally Alaska.

With the decline of herring stocks in the Western Pacific (Japan and Sakhalin Island, Russia), herring reduction plants proliferated in the Pacific Northwest of North America after World War I. From 1920 to 1950, more than one million tons of herring were harvested from Southeast Alaskan waters alone. The peak harvest of 78,749 tons (about one tenth of Japan's peak harvest in 1897) came in 1929 and was processed by 18 reduction plants in the region. The boom did not last, however, and by the early 1930s, Southeast herring were identified as overfished based on stock assessments by biologists (Rounsefell 1930, 1931).

Concerns about declines in spawning herring among the herring industrial fisheries were manifest as early as 1911, when Alaska Oil and Guano company, owners of the flagship herring reduction plant at Killisnoo, testified to a U.S. Senate subcommittee that "During the spawning period, all fishing should be prohibited... enough is known of the habits of this fish to at least inaugurate an effort at regulation"(United States Congress 1912:9). This prohibition was relatively easy to follow for the large herring reduction seiners, as spawning herring generally held less oil, and thus less value, during spawning time. It was more lucrative to fish for sardines and anchovies in California and other lower latitudes until summer, when the boats moved to Southeast.

Martin Perez, a Tlingit fisherman and leader (1918-2014) from Ketchikan remarked on these big 90 foot "sardine seiners."

> Well, [there] used to be a lot of seine boats come up here, great big ones come from Lower 48, you know, in the summer and catch them herring when they're vulnerable, let's put it that way. I know, I remember when those big seine boats used to come in and lay at New England fish dock there, you know.
>
> Our local boats like the Pirate and the Radio and a few of them others were just for the local fishery. Mostly for cold storage you know: bait purposes. But not compared to them great big ones that come from below with ... 90-foot boats. Anywhere from 60-foot on up: big herring seiners. What they were, they were the sardine seiners from California that come up here. That's when the herring started to be depleted (in Thornton, et al 2010a).

Similar protests to those of the Klawock Tlingits were lodged against herring reduction plant seine fishing too close to villages and spawning grounds in Sitka (for Sitka Sound in the 1930s) and Auke Bay (see Chapter 2), as Natives and non-Native locals alike felt the impacts of the commercial take on their community fisheries.

Yet, harvest limitations were not put in place until the early 1940s, after the fishery exhibited signs of collapse. Reduction fishery seine boats continued to harvest remaining schools of herring for another twenty-five years until the last reduction plant closed in 1966. Indigenous consultants who observed the reduction fisheries firsthand agree that this fishery's impacts were significant and long-term and that many local herring populations have yet to fully recover from the exploitations of the reduction era (Thornton et al. 2010a; Thornton and Hebert 2014).

In the meantime, subsistence and bait fisheries have continued, although often in attenuated form as compared to the pre-collapse period. In the late 1970s, a herring sac roe fishery was introduced to supply a growing Japanese market and declining Pacific Eurasian supply of herring sac roe for the *kazunoko* market. The fishery expanded in the 1980s and 1990s and came to be managed according to a guideline harvest level (GHL) of up to 20 percent of the fisheries' biomass (beyond a minimum threshold). This model is based on an age-structured assessment model for herring and measures of egg depositions and observed miles of spawn in a given management area. Although managers consider such guideline harvest levels "conservative," many Alaskans criticize them as not conservative enough. Herring returns have not been consistent under this MSY regime, and ADF&G has had to close many sac roe fisheries for conservation reasons. Herring returns have not been consistent under this MSY regime, and ADF&G has had to close many sac roe fisheries for conservation reasons. At the same time, state managers refuse to recognize herring as a forage fish species, which would subject them to more conservative management. Alaska Native tribes and Canadian First Nations (Jones et al. 2017), including the Sitka Tribe of Alaska, have expressed grave concerns about the MSY biomass and harvest models, the execution of the sac roe fishery, and the impacts it has had on Indigenous subsistence egg fisheries and the marine ecosystem. They have called for more ecosystem-based management.

Today, spawning and rearing habitats throughout the Southeast region remain depleted, although some are showing encouraging signs of recovery. Fishermen we interviewed who plied the waters of central and southern Southeast Alaska in the first half of the twentieth century recall vividly when herring schools extended "for miles" and their feeding activities near the surface resonated like a mighty rain or wind. But such herring abundance has not been experienced since before Alaska Statehood in 1959. By the time the state established modern fisheries management in the 1960s, the baseline had already shifted, and herring were severely depleted in most areas. It appears that under pressure to maximize sustained yield (MSY) in the short term for commercial fishing interests, the state has given much lower priority to long-term planning for enhancement and restoration of regional herring stocks and marine ecosystems.

Sitka Sound Herring in Historical and Cultural Perspective

Even in Sitka Sound, the last stronghold for herring in Southeast Alaska, the actual areas where herring spawn in sufficient quantities for subsistence harvest has decreased by more than 80 percent since the advent of commercial fishing (Thornton and Kitka 2015). Commercial sac roe fishers now compete with subsistence fishers to harvest in the remaining spawning areas, typically close to town. As noted above, this one fishery now accounts for roughly 90 percent of the total regional herring harvest and biomass. All other sac roe fishing areas have been closed or severely restricted due to a lack of sufficient herring biomass to meet the harvest threshold (Appendix 6).

Elders in Sitka see the baseline for spawning populations in Sitka Sound as having shifted dramatically below what they regard as healthy (Thornton and Kitka 2015). In their eyes, the amount of spawn is lower, the quality of spawn is poorer, the duration of spawning has become shorter, and the spawners themselves have become smaller since the advent of commercial fishing, most recently with the ramping up of the sac roe fishery. These changes have had cascading effects on other fish and wildlife.

Sitka Tlingit Elder Herb Didrickson, a salmon troller and herring egg harvester, shared this perspective: "[Y]ou have a lot of trollers who will tell you, 'If we don't have the herring coming in then we won't get the salmon. Herring is the only reason they are here'. Everything in the water feed[s] on herring. We, as subsistence users, depend on the herring. We have the eggs of course, but then when the herring come in and spawn, that brings in the king salmon." Indeed, commercial trollers, recreational and subsistence fishers, and Sitka's Chamber of Commerce were united in opposing commercial herring fishing in greater Sitka Sound during the reduction era (see Craig in Thornton et al 2010a, Appendix B) because they regarded these harvests as harmful to community livelihoods and wellbeing. These large removals also impact the distributional benefits of the fishery to the wider region through gifting and trade.

What is so special about Sitka Sound that makes it an epicenter for herring? Herman Kitka Sr. explained that it is a great mixing and spawning area for herring and salmon. Herring travel in circular patterns, coming from both the southeast and northwest directions.

> The old Tlingits say that the spawning herring comes in from outside.... I done some study with the fur seals and seems like... from Whale Bay up to Point Slocum, these stocks go in a circle... counter-clockwise along the beach and they go back out. Back down [toward Whale Bay].... So all the areas, and I don't know from how far south that ones that spawn that comes up all the way to Crawfish [Inlet], but it's Crawfish. From West Crawfish, Sitka Sound it goes [to] Sitka Sound, [and from] Salisbury, and that's Point Slocum [it comes in from]... and the biggest spawning area is right here, Sitka area. Because they spawn from Dorothy Narrows in here all the way. That's how come everybody from throughout Southeast will come here. Plus, they can time how thick they want their eggs for drying.

According to our oldest informants, who were in their 90s in 2008 (e.g., Herman Kitka, Sr., and Walter Soboleff), places like Kootznahoo Inlet, Seymour Canal, and Sitka Sound felt the impacts of reduction fishing as early as the 1920s and 1930s. The Killisnoo reduction plant, as the only one in the region from 1882 to 1907, disproportionally affected the Angoon-Killisnoo region. According to Walter Soboleff, who worked at the Killisnoo plant in the early 1920s, fishers supplying this plant targeted herring locally until supplies were diminished, at which point they moved down Chatham Strait and into areas of Frederick Sound (Thornton et al 2010a).

By the 1930s, Sitka Sound—by far the most productive herring fishery in the region—was also being impacted to the extent that local organizations, both Native and non-Native, were calling for a prohibition on herring seining. Such restrictions helped the Sitka Sound herring population rebound over time. The largest regional impacts of the reduction fishery took place between 1920 and 1960, when herring seiners increasingly targeted more remote areas of Southeast Alaska, predominately lower Chatham Strait, Frederick Sound, and southern Baranof Island, which were less visible to local communities. According to Herman Kitka, Sr., overfishing in Larch Bay (southern Baranof) was particularly damaging because those herring were part of the southern migration of herring that spawned in Sitka Sound.

The cumulative impacts of these removals became increasingly apparent to fishermen plying these waters, who had been used to seeing Chatham Strait "black with herring" (Skidmore 1885) and visible, as many Elders put it to us, "as far as the eye could see," and their roiling surfacing activities audible "like rain" or "wind" until at least the 1930s. Soon the quantities of herring perceived and caught began to decline in areas which had once been superabundant or seemingly limitless.

During this period, after the peak harvest of 1929, the cautionary reports of biologists such as Rounsefell were largely ignored. Yet, concern for herring conservation was raised in the U.S. Congress as early as 1939, when a fisheries manager proclaimed that commercial herring fishing in Alaska had been "in the past too extensive" (United States Congress 1939:646). By the time statehood was achieved in 1959 and a modern fisheries management regime was put in place in the 1960s, herring were already depleted in many areas, although just how much is difficult to quantify.

The Sitka Sound commercial sac roe seine fishery commenced in the 1970s. The purse seine fishery nets herring (both males and females) by the thousands of tons at the cusp of spawning in order to harvest the roe skeins from pregnant females for export to Japan and eventual production into salted roe, or *kazunoko*, for Japanese consumers. The fishery was born to supplement the commercial bait and emerging spawn on kelp fisheries, and especially to serve expanding consumer markets for *kazunoko*. With Japan's economic rise and the island nation's collapsed herring fisheries, local supplies could not meet this market demand due to previous overfishing. This market peaked in the mid-1990s in terms of ex-vessel value

(Appendix 1, 1996), coinciding with Japan's economic ascent during that period, and has not reached the same heights since.

Fishing quotas in Sitka Sound, on the other hand, have risen steadily, climbing to more than 28,000 tons in 2012, fifty times those set in 1975, when the first herring population assessments were carried out by the ADF&G (Appendix 1). Due to growing biomass estimates, Sitka Sound sac roe harvest quotas have increased from less than 800 tons in 1976 to a record of 29,000 tons in 2012. The market for sac roe, however, is not growing, as Japanese cultural tastes have shifted away from *kazunoko*, causing prices to stagnate and decline. Delivering more supply does not solve the problem unless markets can be found for the non-roe portion of herring. In fact, sales of herring carcasses—waste from sac roe production—have increased but are not highly profitable and create few local jobs. Most carcasses are shipped to China for processing into animal feed, including food for farmed fish (such as Asian carp).

The framing of herring as a commodity, harvestable according to the principles of MSY, has cascading consequences, not only for herring populations but also for other actors in the marine ecosystem. It underplays the mobility and agency of the fish and ignores the role of other dynamic ecological constraints on the productivity of fish populations, such as predator-prey relations, climate change, and habitat conditions. Yet we know that these broader ecological factors are linked not only to the status of herring populations each year, but also to the resilience of social-ecological ecosystems over time (Berkes 2003; Walker and Salt 2006).

While lacking the "charisma" and anthropomorphic qualities of larger fauna in the Western imagination, Pacific herring are nevertheless regarded as a prestigious keystone species by Tlingit and Haida. As Mike Miller of Sitka put it, "I think [herring are]... really key to the different populations of fish here... compromise of the overall biomass is definitely going to have a big effect on the salmon fisheries and the halibut fisheries and the marine mammals... it's the key to the ocean... it's our buffalo."

Framed in this way, herring's annual returns to particular spawning grounds at Sitka Sound and other areas were considered voluntary and responsive to human cultivation behavior within a moral ecology of respectful relations (Thornton and Kitka 2010). Thus, when cultivating spawning areas with western hemlock boughs for herring to spawn upon, one took care to make the substrate and its surroundings attractive, or pleasing, to spawning fish, and even to talk to them or their spirits directly (Thornton et al. 2010a; Thornton 2015). This meant laying the boughs carefully in the intertidal zone and minimizing disturbances, such as loud noises or agitation of the waters around spawning beds. This is because herring reassess their spawning areas each year, even if they have spawned in areas previously. This initial period of scouting by older male herring is very critical, as perturbations of the environment can interfere with spawning and distress the herring, causing them to spawn too early or to move to a less optimal location. The Indigenous protocol is to wait for the older herring to complete this scouting for spawning areas, whereupon the older males will release pre-spawn milt (also called false spawn) to attract females. Only then is it appropriate to place branches in the water, at least in sensitive (often tightly enclosed) areas. Historically, "Watchmen" actively patrolled spawning areas to prevent threats to this sensitive scouting and initiation of spawning period, to insure renewal of the herring life cycle (Thornton et al. 2010a). These beliefs and practices, and the integrated knowledge and skills that underpin them suggest that Tlingit and Haida peoples not only were aware of potential human impacts on local herring spawning, but also of their responsibilities to maintain respectful relations to safeguard continued herring returns. It was all part of what Berkes (2012) calls a knowledge-practice-belief system, evolved within particular social-ecological contexts, and especially at Sitka Sound.

Recent scientific studies support the Indigenous knowledge that older fish "entrain" younger fish into spawning areas (Petitgas et al. 2006). Modeling this "go with old fish" behavior (MacCall et al. 2018) indicates that it may help to maintain a higher diversity of spawning areas because older fish possess stronger learned fidelity, or migration memory, to their "home" spawning ranges. Thus, when high proportions of older fish are removed from spawning schools, as may happen in a sac roe or bait fishery, it will eventually reduce the diversity of spawning areas used. Without as many older herring to lead them, younger herring will simply follow the crowd to the larger spawning areas. This could explain what is happening in Sitka Sound; with the decline of smaller spawning areas throughout Southeast Alaska, Sitka has become the herring spawning "mecca" for more than 90 percent of the spawning population. Not coincidentally, this shift occurred in the 1990s when Sitka Natives also noticed that their "spawners" were getting smaller (more immature) and hence less likely to have been entrained by older fish. Although it can be difficult to distinguish between spawning behaviors associated with reduced abundance and those associated with fishery impacts on age composition, Tlingit and Haida Elders consistently emphasized the important "leading" role that older fish play in showing younger fish where and when to spawn.

Similarly, Tlingits and Haidas are mindful of other environmental contingencies that affect spawning, such as seawater temperature, wind conditions, large spring tides, and the diverse quality of the spawning habitat. For example, the "washing of the spawn wind," or *L'uk' eeti.oosk* in Tlingit, helps break up and distribute herring eggs, thus contributing to greater hatching and survival. Through this relational epistemology (Bird-David 1999), Sitka Tlingit herring egg harvesters actively cultivated new spawning habitat with substrate to increase the abundance, distribution, and predictability of spawn (Emmons and de Laguna 1991; Thornton et al. 2010a).

Yaaw Teiyí (Herring Rock) and the Sacred Origins of Spawn Cultivation

The cultural and spiritual interdependence of the herring and Natives of Southeast Alaska is exemplified in the story of Herring Rock, encapsulated earlier, but related here by Kiks.ádi leader Louise Brady in her testimony to the Board of Fisheries in 1997:

> It all [the Kiks.ádi sacred relationship or covenant with herring] happened with a young lady who each morning would go down to the beach and sit on the rock, and then all the herrings would come around the rocks, swimming around. And then she would start singing lullabies to the herring. Each morning as the tide came up, she'd go down to the rock and the herrings would come swimming around her, and the waves were dancing around her. Then they would disappear again. Then a few days, a few

> weeks later they came again, then they spawned around that place. From then on it spread all through the village, all the way to Jamestown Bay, out past where the Park Service is, and then all among the islands. And therefore it became known as the Herring Rock (Brady 1997).

As the story suggests, the herring came to Herring Rock because of the presence of the Kiks.ádi woman who cultivated them with song, and due to the attractiveness of her long hair dangling in the water, which lured the herring to spawn on it as substrate. This human-herring relationship was consecrated through mutual respect, engagement, and interspecies communication. Through the return of the two parties annually to renew this covenant at *Yaaw Teiyí*, or Herring Rock, the site became the first and main locus of spawning in Sitka Sound.

Herring Rock, the story behind it, the images or crests representing it that adorn Kiks.ádi regalia, such as the Herring Rock Robe (*Yaaw Teiyí Naaxein*, see Thornton 2008:175), and the related dances, songs, and names that come from these events, such as *Kaxátjaa Sháa* ("Flipping Girls," given to Kiks.ádi women), are all *at.óow*, or sacred possessions of the clan. Thus, the Tlingit Kiks.ádi relationship to *Yaaw Teiyí* is not simply recounted as a myth and claim to Herring Rock, but also is enacted as a covenant and charter for contemporary interactions with herring, based on respect, reciprocity, and sustainability.

The charter represents the "herring-as-agent" model of ecological thinking, which is at odds with the dominant "herring-as-commodity" perspective and is therefore dismissed. From a Tlingit and Haida perspective, disrespect for herring endures under the current MSY exploitative regime. What is more, the commodity model has definite geographic or spatial effects: it homogenizes ("deterritorializes") space by rendering out critical relations that compose the Indigenous space—particularly Tlingit social and sacred relations—of which herring are a primary agent. This disrespect is perpetrated through the narrow economic framing of herring as an MSY commodity and through the complete marginalization and subordination of the Indigenous knowledge and relational and reciprocal values toward herring.

The Sitka Herring Festival, recently revived by the Sitka Tribe of Alaska as a means of countering the reductionist model and commodification

Nels Lawson (center) explains the significance of Herring Rock and the herring season: "The Herring Rock has a very long history among our people. Very significant. The days of long ago, the herring would come to our homeland and the first place the herring would spawn would be at the Herring Rock and this time of year was very significant to our people, the arrival of the herring on our shores that meant the arrival of new food, new food for our homes and the end of the old foods that sustained us through the winter months" (http://www.bethany-goodrich.com/blog/2016/8/chasing-herring-spawn-on-sitka-sound). Kiks.ádi leaders Fred Hope (left) and John Duncan (right) look on. (Bethany Sonsini Goodrich, Sustainable Southeast Partnership).

logic surrounding herring, may help to recast herring in a more social-ecological frame. The festival celebrates the importance of herring as a cultural keystone, biological foundation, and ecosystem umbrella species. The festival also seeks to bring everyone in the community together for cultural events and science education in a multi-generational celebration of herring, honoring the multiple roles the fish play in ecosystems, cultures, and economies. Sitka has always attracted visitors during the herring spawn; it is a time for people to come together to celebrate the abundance and importance of herring, their resilience in returning each year, and the coming of spring. The festival was initiated by the Sitka Tribe for the benefit of the whole community and region and includes the blessing of the herring at "Herring Rock" as a highlight. The blessing ritual is a communal rite designed to bring attention to the importance of a healthy herring population to Sitka Sound's social-ecological system, precisely because of the distribution of benefits that flow from their abundance.

Even though Herring Rock now stands in front of the *Sheet'ka K̲wáan Naa Kahídi* (Sitka Tribal Community House), having been partially destroyed and removed with waterfront development in the early 1970s, the blessing ceremony persists. The late Donald "Duck" Didrickson, a Kiks.ádi Elder who led the ceremony for many years, observed:

> It had three [English] names you know; it's Didrickson Rock, after my Grandmother, Chris Didrickson. That's who owned the Herring House. That's why it's called the Herring House [*Yaaw Hít* or *Kaxátjaa Hít*]. The reason they call it Didrickson Rock— everybody give up [on it]. They couldn't get any herring spawn [there]. We set branches down there on the beach and all the herring come in. I figure it had the power to bring in the— see it had something to do with my people settling this area.... But Herring Rock... had the power to draw everything in, everything we wanted, you know.... Draw the herring and everything else: the salmon, the halibut, the seaweed, gumboots, everything. That was our— in the old days, that was our calendar. Herring spawned down there, you know, spring is sprung.... There's a lot of history to it you know... Yeah, Herring Rock been there over 10,000 years.... And they always figured it had the power to attract the herring.

The blessing, attended by fishermen and members of the Native and non-Native community, is a public means of honoring the importance of herring and emphasizing the need for respect and restraint in caring for them. The ceremony also reaffirms the shared belief in the power of cultivated habitats like Herring Rock to draw herring in to spawn on substrate—hemlock trees and boughs—placed by their Tlingit hosts.

Kiks.ádi Elder John Duncan notes that the placing of boughs at Herring Rock traditionally was accompanied by an invocation to the male herring who probed the branches place around the rock before releasing their milt. As the hemlock set was laid in the water, the speaker would say in Tlingit, "*Yee Shaatk'í jeeyís áyá!*" ("This is for your (all's) [mature male herrings'] girlfriends [to spawn upon]!", see Thornton 2015).

Conclusion: Ecosystem Stewardship and Distributional Benefits

Until recently, the exploitation of small, nutritious forage fish like herring has received little sustained attention in contrast to declines of more charismatic and individually-valued commercial fish, like blue fin tuna, swordfish, cod, and salmon (Alder et al. 2008). Yet, as Tlingit and Haida ecological knowledge has long maintained, herring is the "feed for everything," the key to marine ecosystem productivity, health, and sustainability. Herring must be cultivated sustainably, rather than exploited maximally, in order to remain productive for human populations and other species that rely them.

Key to this concept of sustainability is recognizing the deep and wide array of distributional benefits tied to herring's immediate food value, its temporal exchange value as stored nutrition and cultural capital, and, most importantly, its social exchange value in connecting people both within and across communities throughout the Southeast region and beyond via sharing, gifting, barter, and customary trade. These benefits are physical, economic, social, and cultural, and integral to subsistence lifeways, which support Southeast Alaska Native cultures, as articulated in the ANILCA federal subsistence law.

Overexploitation of herring can wreak long-lasting damage to the fisheries themselves and the marine food chains they support (Smith et al 2011). In a recent assessment of prospects for rebuilding global fisheries, Worm et al. (2009:584) argued that MSY "should be reinterpreted as an upper limit rather than a management target" and overall exploitation rates reduced to improve the health of marine ecosystems. While this is reform in the right direction, it should be noted that the authors' vision of "a new cooperation of fisheries scientists and conservation biologists sharing the best available data" all too characteristically neglects local and traditional knowledge (Worm et al. 2009:584). In Southeast Alaska, the Indigenous knowledge, stewardship practices, and multi-species communal values that allowed marine ecosystems to thrive on a superabundance of herring remain neglected, if not invisible, to present day fisheries managers who regard their MSY model as optimal.

There was a time when herring were so plentiful that Indigenous inhabitants could rake them easily from the ocean's surface with spiked paddles. The reduction fishing era undermined this richness by fishing down the herring superabundance to scarcity. More careful management of fisheries in the last forty years has enabled herring to recover but only to a fraction of their pre-industrial fishing levels. Today, Southeast Alaskan herring are managed in a depressed state now considered under MSY as "normal," and largely for a global sac roe fishery which benefits a limited luxury food market in Japan and a few dozen commercial permit holders in America. Simultaneously, local subsistence herring egg gatherers and others dependent on herring are experiencing the deleterious effects of shifted baselines and management frames. Policies of MSY married to neoliberalism may avoid tragedies of the commons, but they do not prevent tragedies of commodification resulting from overexploitation and alienation of biological and cultural keystone species in marine ecosystems cultivated by aboriginal peoples over millennia.

Despite its citizen participation and scientific management orientation, Alaska has lagged in its development of a place-based ecosystem-based approach into fisheries management. Instead, as Carothers and Chambers (2012:51) note, "[m]any fishermen and fishing communities have been alienated from fishing livelihoods through processes of commodification and marketization of fishery access rights." The lack of strong historical-ecological frames, respect for local knowledge of herring and their benefit flows, and the State's political marginalization of Indigenous knowledge and subsistence economies has made State fisheries managers both reluctant and unable to fully recognize the importance of abundant herring to cultural and ecological wellbeing. This marginalization has engendered increasing resistance and calls for a change in management toward a more balanced and broader valuing of herring in marine ecosystems.

Tlingit and Haida peoples have intimate knowledge, experience, and techniques in place-based marine ecosystem stewardship, which have proven effective over time and may be useful as adaptive management tools (Berkes et al. 2000). These tools, including habitat protection, cultivation and enhancement techniques, and herring transplantation and harvest management strategies, could become the basis for stronger ecosystem management, habitat revitalization, and sustainable fisheries in the future (Thornton et al. 2010a; Thornton 2015). Significantly, the Haida Nation of Canada is pursuing such an agenda based on comparable experiences with their herring stocks under federal management (Jones 2007). The Haida, along with other West Coast First Nations, sought and won an injunction to prevent the Department of Fisheries and Oceans Canada from opening commercial sac roe fishing in areas of traditional importance to their marine ecosystems (Thomas 2014).

In the meantime, notwithstanding their resilience and mind-boggling ability to survive, it is not clear if the vast waters of the North Pacific rim, once teeming with herring, will suffer further local disappearances of this critical species, compromising the regional biodiversity and other capacities they support, creating a serious reduction in the benefits that have historically flowed through Indigenous subsistence communities with herring production, protection, and distribution.

Many Tlingits and Haidas fear this will be the fate for Sitka, the last true stronghold for herring in Southeast Alaska. Native leaders are working across boundaries to prevent this. As Kiks.ádi Elder Louise Brady put in a recent post on the "Herring Rock Protectors" (founded in 2018) Facebook page, concerning the 2018 Board of Fisheries meeting in Sitka on herring:

> I testified before the Board of Fish in 1997 because the Herring stocks were seriously declining. Over the last year I have listened to many friends of mine who grew up here, fishing and gathering eggs for our families. They ALL say... the herring are disappearing. Many people did not get herring eggs last year... or the year before. It was daunting to testify before the Board of Fish in 1997. I'm sure it won't be a picnic next week. But I want to speak for what is right. I want my grandchildren to enjoy the

> same foods I have enjoyed here on Sheet'ka! So I will testify. If you don't want to testify, please come stand with us. If you are Alaska Native or not. Please join us. If you want me to sit with you when you testify I will. This place... Sheet'ka... Sitka... is special. If we stand together... we can keep our way of life and continue to enjoy this special place. If we don't... it could be another:
>
> - Auke Bay-no Herring
> - Tenakee Inlet
> - Hobart Bay
> - West Behm Canal
> - Ka[h] Shakes

The Sitka Tribe's position has been attracting supporters, both from the local Fish and Game Advisory Council and the City and Borough of Sitka Assembly, which passed a resolution in January 2018 declaring that it "shares the Sitka Tribe of Alaska's concerns about conserving the Sitka Sound herring stock and protecting the subsistence harvest of herring eggs in Sitka Sound" (City and Borough of Sitka 2018; see Appendix 8).

At the Alaska Board of Fisheries meeting in February 2018 in Sitka, STA proposed capping the commercial herring harvest at 10 percent of the estimated biomass (versus today's sliding 12-20 percent scale) beyond a minimum threshold of 25,000 tons of herring (versus the current 20,000). However, the board took no action on the proposal, choosing to preserve the status quo. The fishery failed to meet its quota by a large margin and had to be closed early for the fourth time in the last six years. In the aftermath, subsistence herring harvesters had one of their worst years on record, with many failing to get enough eggs to distribute. This has only accentuated concern that Sitka Sound may lose its herring stocks as a result of poor management, as Elders had been predicting would happen for decades, if the management model did not change (Thornton et al 2010a).

On December 11, 2018, the Sitka Tribe of Alaska filed a lawsuit against the Alaska Department of Fish and Game and the Alaska Board of Fisheries in Superior Court, charging them with unlawful conduct and requesting an injunction against ADF&G that would require them

Louise Brady (center) of the Kiks.ádi clan wears the Ravenstail Herring Robe, woven by the late Teri Rofkar, to perform the herring dance at a *ku.éex'* or "invite ceremony" held on January 14, 2018, and sponsored by the Herring Rock Water Protectors, a group concerned about the failures of ADF&G herring fisheries management to support herring and herring subsistence uses. Elsa Sebastian (left) and Tiffany Justice (right) accompany her. (Photo: Bethany Goodrich, Sustainable Southeast Partnership).

to develop a new management plan for the fishery. In a press release, STA chair Kathy Hope Erickson called for respect for the benefit flows enabled by Sitka Tribe members cultural responsibilities to and uses of herring. "The time is now," she said "to ensure our people have the chance to fulfill their cultural responsibilities which have been interwoven with the herring since time immemorial, and to fill their freezers. We cannot sit by while the State of Alaska shirks its statutory and constitutional duties to citizens. We demand action by the state."

This means shifting away from the current MSY model of commercial herring sac roe production, which has imperiled subsistence uses at Sitka Sound, towards a MSB (Maximum Sustainable Benefit) model, which recognizes the wide ranging benefits to Alaskans of the Native distribution of herring eggs.

CHAPTER 3: HISTORICAL HERRING EGG DISTRIBUTION AND TYPES OF EXCHANGE

"What does [sharing herring roe] mean?
That means that they think of you.
That means they love you enough to share."

—Heather Powell, Hoonah

Photo: Bethany Sonsini Goodrich, Sustainable Southeast Partnership.

In the previous chapters, the general characteristics of the herring fisheries were examined, including the modern context of management and conflict between commercial and subsistence production. It was shown that Euro-American and state management systems have operated from conflicting cultural models of herring. The state's model seeks to maximize sustained exploitation of herring as a resource and commodity. The Southeast Alaska Native model, in contrast, has traditionally operated according to a covenant of respectful engagement with herring to cultivate successful spawning while at the same ensuring sufficient eggs for consumption, sharing, and exchange.

This chapter discusses the development of the historical herring distribution and exchange system specifically, including levels of participation and other key patterns and attributes of the distribution process itself.

Levels of Participation in the Production and Distribution of Herring Eggs

Levels of participation in herring egg harvesting were higher in proportion to the population in the past than today, while historic levels of distribution have not been estimated comprehensively, despite the early accounts attesting to copious amounts of herring eggs being transported out of Sitka Sound to other communities. The German explorer Aurel Krause (1956 [1885]:39), who traveled to Alaska in the 1880s, wrote a historical sketch of the area, citing reports by the Russian colony administrator Murawief from 1821-1826, which reported "In the spring…

about one thousand people assembled there (Sitka Sound, Baranof Island shoreline) and perhaps as many again on the neighboring islands to gather eggs." Based on this estimate of active gatherers, if we consider 2,000 people present just within Sitka Sound, this might equate to about 500 households or production units. If each household or family unit harvested on average 1,000 pounds of herring eggs (or 250 pounds per capita, though obviously not all were harvesting), that would translate into 500,000 pounds of herring eggs for a Tlingit population considerably smaller than today. This figure is more than double the "amount necessary for subsistence" (ANS, 136,000 to 227,000 pounds,) as determined by the Alaska Board of Fisheries presently. If participation levels had remained constant, and using the same conversion factors, today the Sitka Tribe, boasting a population of more than 4,000 (1,000 households x 1,000 pounds per household), might need closer to 1 million pounds of herring eggs. Today, harvesting has become concentrated among "super households" that have the labor and means to harvest efficiently (Schroeder and Kookesh 1990), but this has not necessarily lead to drop in demand. Moreover, this figure does not include harvesters coming to Sitka from other communities, of which there are at least dozens.

Survey results from ADF&G, however, estimate that household participation rates in the subsistence herring egg fishery are closer to 100 households, even incorporating non-local harvesters. This finding suggests a 90 percent drop in participation over the past century or so. Why? According to interviewees, the reason for lower levels of participation are directly related to the opportunity costs of pursuing egg harvesting. The current ADF&G management regime has led to reduced spawning extensity, duration, predictability, and quality, in turn leading to increased competition and material and sociocultural costs to find and capture suitable spawn, either on Western hemlock branches (placed as substrate) or kelp (naturally occurring hair or macrocystis kelp). Areas once proximal and reliable are no longer supporting herring spawn or producing an unsatisfactory quality of spawn for subsistence uses. With fewer areas of spawn there also is increased jostling among harvesters for good setting spots, including from industry-supported boats that are paid to harvest eggs by the sac roe fishing advocacy organization (Sitka Herring Conservation Alliance). There is also increased need for vigilance, due to theft or misidentification of sets in the crowded seascape, which adds more time and fuel costs to harvesters seeking to avoid losses.

As a result, it takes more effort to get herring eggs than it did in the past. Numerous egg harvesters reported that they must spend more time monitoring the spawning conditions, scouting for spawning areas, finding suitable setting sites, setting enough trees (with poorer and less predictable spawn) to gain an adequate harvest, and protecting their hemlock sets from disturbance and theft. Aaron Bean, a modest but consistent harvester in his 40s, estimates that he must spend "10 times" as many hours as he used to when he was young to get the same amount of eggs, due to these factors, with the main costs being having to cut more trees and make more sets due to the unpredictability of spawning and lack of return on many trees (little or no spawn, theft, etc.). This represents a significant time cost differential, and the return on additional time investment is not guaranteed, due to the volatility of herring spawning in Sitka Sound in recent years.

At the same time, the value of other choices not taken in favor of pursuing herring eggs has risen. Consider, for example, the fact that participants may have to take time off from work, giving up valuable wages, or to neglect parental or other duties, in order to secure herring eggs, or the fact that the cost of fuel and search time for herring spawning areas has increased. These opportunity costs render herring egg harvesting too expensive for a significant number of would-be participants.

These costs are magnified for out-of-town herring egg harvesters, who must travel farther, adding even more time and expense. The effect of these rising opportunity costs has diminished non-local participation, although dozens of non-local Natives, especially those with ties to the community or support from their own communities, do come to participate in the fishery. Others seek to coordinate their visits to family or services in Sitka (e.g., SEARHC medical) to coincide with herring egg time, so they may participate as recipients in the Sitka Sound distribution.

The earliest accounts from the Russian period, cited above (Schroeder and Kookesh 1990:4), report that "Some [Sitka] Elders recalled that visiting clan members would stay at island camps near Sitka that were used by local branches of that clan. Informants agreed that the Sitka Sound herring egg harvest was open to members of other communities who were free to harvest eggs where and in whatever quantity they wished." This suggests that cooperation rather than competition governed the participation of non-locals in the fishery. The extensity of harvesting areas also meant there was little need for territoriality or boundary defense because the areas accessible for setting branches and harvesting spawn on kelp were abundant and extensive, unlike the situation today. This is consistent with Dyson-Hudson and Smith's (1979) "economic defendability" hypothesis, which suggests that resources otherwise dense and predictable will not be territorialized in the case of superabundance. Interestingly, even with the contraction of extensity and abundance, territoriality has not developed (cf. Acheson 1988), except to the extent of individuals defending their particular sets from theft. This lack of territoriality may be due to in part to the lack of legal support for territorial partitioning (except into subsistence only and non-subsistence fishing zones) but also to the increasing lack of predictability and rising costs of defending a harvesting area in the contemporary spawning environment. However, a counter explanation is that with the high expectations for distribution, whoever is successful in harvesting will be distributing most of their eggs anyway, including even potentially to friends and relatives of any competitors—if not to the competing harvesters themselves. In this way the camaraderie of the fishery still adheres despite the rising competition for good spawning sites.

Patterns and Attributes in the Distribution of Herring Eggs

Distribution can be described using a number of analytical attributes, all of which are important to understanding the nature of the herring egg production and distribution systems.

Extensity—what is the geographic extent of herring egg distribution? The geographic extent of herring egg distribution has been generally limited to the Western North America. This includes all the Western states, the Canadian province of British Columbia and the Yukon Territory, and all major regions of Alaska. Interestingly this trading sphere is not that different from the traditional trade area of the Tlingit, who ventured north, south, and inland throughout much of the Western U.S. by canoe or small boat, though not as far inland—e.g. Sandpoint, Idaho, or Washington DC—or into the Pacific—e.g., Hawaii and even the Philippines—as some eggs travel today by air.

Intensity—what quantities of herring eggs are distributed within defined periods and recipient communities? The intensity of distribution by area has never been consistently measured. Generally speaking, however, the intensity of sharing distributions may be dictated by social ties. This explains the high degree of sharing of eggs that takes place between Sitka and Hoonah and Sitka and Angoon, for example, where familial and clan ties are strong. This is also true of the trade between Sitka and Juneau, despite the longer distance; Juneau's status as a regional center means that Sitka Natives also tend have relatives who have migrated there.

Trade distribution also is a function of comparative advantage. For example, if Sitka has spawning herring but no moose, and an interior community, such as Haines Junction, Yukon, has plentiful moose but no spawning herring, each community is theoretically incentivized to capitalize on its comparative advantage in producing its distinctive food and to trade for that which it doesn't have (moose in the case of Sitka; herring eggs for Haines Junction). This pattern still adheres in many cases, but trade distribution patterns, both historically and in the contemporary era, have involved a variety of forms of sharing and exchange, which may be defined according to the nature of the reciprocity, as follows.

Generalized Reciprocity—is gift giving without an expectation of an immediate return. It generally develops as a kind of social insurance among societies where activities such as hunting or herring egg gathering may depend partly on luck (picking the right spot, etc.), and thus by giving away when you have success, you can expect others to do the same when you are unlucky and in need. The expectation to give is generalized as opposed to calculated, based upon what one might receive in return. For the Tlingits, the nature of generalized reciprocity is closely tied to capacities. An able harvester generally will try to harvest and provide eggs to those who cannot get them for themselves (due to failed effort, lack of capacity, etc.), and expects the same in return, say, for example, when the harvester becomes too old or ill to go out himself (or herself). Sitka Elder Nels Lawson explains: "I never charged anybody. And whatever it took to get them their herring eggs, that's what it took" (NL). Now, when

Nels receives eggs each year, he doesn't feel obligated to reciprocate: "It's kind of—going from when I was much younger I gave out a lot. Now it's coming back." This is the essence of the principle of generalized reciprocity. The exchange or return on giving may be quite delayed, or not come at all. As Tammy Young of STA observes:

> "[I]n the end [it's] not that anybody really expects anything back, but you know you send herring eggs up to Barrow, you might get some muktuk, you know. I mean, you send up to Nome and you might get some you know dried fish or...yeah. So, it's—herring is a very interesting vehicle. It allows so much life to happen through it." (TY)

This observation is important, because it relates to the fact that in a society defined by reciprocity, the act of giving is a vehicle—a vehicle of communication between parties who are related or otherwise care for one another. As Tammy Young suggests, "These are loving reciprocities that just naturally happen .... So it's not like anybody plans it, but yeah, I think that that...those relationships are strengthened over the course of time by that interaction."

What's more, a generalized reciprocity system promotes generosity and a sense of mutual dependency and connectedness which is vital to perpetuation of the culture itself. As Jerrod Galanin explains:

> "But thankfully, my Uncle kind of taught me in his own way that it's not always about what you can immediately get. Like, you do something, it doesn't mean you're going to pay me, you're going to get something. You're going to get something out of it, like for yourself. And that's the same with going out and gathering. You're already out there, we're doing it. I mean, it feels good to be—to help people out, you tend to perpetuate I guess, just to keep this going; keep the sense of community and sense of culture." (JG)

Chuck Miller elaborates on how under a generalized reciprocity system, the goodwill generally "comes around":

> "I gave away herring eggs one year to somebody and then that Christmas, or before Christmas, or somewhere around theere between Thanksgiving and Christmas—I think it was before Christmas—I got a bunch of jams and jellies from that person. I wasn't worried about it, but it was so nice when I got them. I'm like, 'Oh man! Homemade jams and jellies? Yeah, twist my arm.' But it comes around. And if it doesn't that year, maybe the next year. Or they'll say—or they'll do it something way down the line. It's just something that's been handed down to me and it's been handed down to me freely; why not. Just, share freely." (CM)

RH, a retired seine boat captain from Hoonah, put it succinctly when asked why he would take his boat all the way from Hoonah to Sitka to get eggs just to bring back and distribute. "You do it because you want to be a generous person...which is important among our people." In other words, the expectation is to share if you can, not only as a kind of insurance policy to support those who otherwise would go without and who will in turn support you when they have eggs so you will not go without, but also as a means of building respect and prestige as an ideal social person (cf. Kan 1989, Thornton 2008).

Balanced reciprocity—is defined as comparable return on gifting within a balanced timeframe. This is typified by what happens in a gift exchange among friends. Be it birthday gifts or Christmas presents, friends often exchange gifts of approximately the same value, on approximately the same timescale (annually in the case of birthdays and Christmas). Most barter and trade are defined by balanced reciprocity; one may offer a bag of dried seaweed in exchange for a jar of smoked cockles, for example, and regard this as a balanced exchange with each participant finding it "worth it" to broaden their portfolio of foods. The sense of balance is not fixed but relational between the exchangers. Schroeder and Kookesh (1990:7) recorded a number of general phrases for exchange, including: *aat yax* (general exchange); *aa daa seix* (*a daséix'* or *woosh daséix'*, exchange)—noting that "If a person feels it is not fair he says: *aat dei akwstaak* (*aadé k̲k̲wast'áak̲*, "I will add more to it"), and *aa yeen dax* ("Split right in half" [or evenly between])." Elders stated to them that "Tlingits were always fair to each other in order to always protect good will." Indeed, *wooch yáx̱* (balance) is identified as a core Native value by Sealaska, the regional Native corporation (https://www.sealaska.com/who-we-are).

Mainland Natives on eulachon rivers often trade one or more jars of eulachon grease/oil for a box of herring eggs, because it feels like a balanced exchange, with each regarding the opposite's product as a luxury or delicacy. However, in cases where a simple food-for-food exchange is not possible, other arrangements can be made to balance the transaction. Sitka Luknax̱.ádi Clan House Leader Chuck Miller, who has harvested herring eggs in Sitka for many years, explains:

> "[I]f people wanted it [herring eggs] out of town, they would make arrangements with me ahead of time and either barter with me, or I would just flat out say, 'Ok, well, if you would like some, I don't mind getting it. You just need to pay for the wet box and you have to pay for the shipping because I am not doing that.' They're pretty good about throwing in some gas money. They'll just send a check or something. I don't ask for it, they just do it. Or sometimes people will barter like a box of, like sometimes I've received a box of dried halibut or seaweed and maybe seal oil" (CM).

All of these transactions, including the reciprocation of gas or shipping money for eggs, can be seen in the context of balanced reciprocity, although the balance may be tempered by socioeconomic factors, such as status or ability to pay. For example, Schroeder and Kookesh (1990:7) recorded the phenomenon of the "love gift" (*du kaìani di gee du* "material thing from a relative"), which is "prompted by the receiver" (i.e., a loved one) and involves giving less than what might be considered the fair exchange value of the item, with the giver accepting the transaction because of the person's status (as a relative, etc.). The implication is that the imbalance may be mitigated by the status of the recipient or equalized over longer time frames.

Many Native Elders are pleased to receive some herring spawn on kelp from STA's distribution program or other sources. (Photo: Bethany Sonsini Goodrich, Sustainable Southeast Partnership).

When balanced reciprocity is achieved without negotiation, it is considered respectful. Roby Littlefield elaborates on how balanced reciprocity works, ideally:

> "I do appreciate the respect of getting a gift back. Sometimes the gift is in the form of money, sometimes the gift is in the form of barter and trade. Like every year until this year I was able to trade with Yakutat for red seaweed.

> So [one woman], I sent her a 50-pound box of herring eggs and she sent me a great big box of completely dried and ready to go *k'áach'*, that's the red ribbons [seaweed]. And that lasted me, I could share it, it was enough that I could share it with people around here." (RL)
>
> "I've sent herring eggs to Whitehorse, and six months later they're coming into town and they bring me a box of moose meat. So, it's not discussed, and when they ask how much I'll just tell them well let's trade something, just send me something. And it's on their own honor whether it's what value [i.e., what value it has]. And actually, respect is a source of value in our culture. It's worth more than money. And if somebody gives me respect I'm very happy to not get any money or anything else. Respect is critical in trade and barter." (RL)

Balanced reciprocity may also be achieved by showing respect through exchange of non-food items. David Kanosh, a recipient of herring eggs who is no longer able to harvest himself, explained some of the gifts he gives in exchange for herring eggs from family or friends, or even STA:

> "And then sometimes you know, just in kind, … I'll give a little something back to them, like I'll bring chai, Hudson Bay tea or you know, devil's club just as a 'thank you,' which they don't really expect but you know, it's what I'll do…It's a kindness that … we pay to each other. You know, not necessarily monetarily but being able to sit down, share a cup of tea and … talk about things in general." (DK)

Balanced reciprocity also can be facilitated through formal or informal trade partnerships. This is an old practice especially between adjacent cultural groups. Traditionally, partnerships existed through individual, house, or clan lines, but might also be generalized to the community or ethnic level. A formalized trade partnership could be based on real or fictive kin ties. In the case of coastal Tlingit and Inland Tagish people, for example: "They had to have partners, Tlingit partners and the Tagish people. They traded down here (Tagish)" (Annie Auston in Thornton 2004:95). Preferably, a trade partner would belong to the same moiety if not the same clan in Tlingit and would be called *ax̱ yak̲áawu* (translated by Olson [1936] as "my own man" and by McClellan [1975:506] as "my respected friend") or *woos' yak̲áawu*, "mutually joined together" (Emmons 1991:57). The partnerships themselves might be consecrated with gifts (Olson 1936), and in some cases would be further cemented through marriages (see Olson 1936; Thornton 2004: 89-99). This was particularly prevalent in the case of trade with interior tribes, such as Tagish and Southern Tutchone, in part because the trade volume became so significant that it was necessary to formalize the partnerships and institutionalize the trade.

Tlingit trade expeditions could be large and long-term affairs. As Oberg reports (1973:105-106):

> These trading expeditions were of no mean proportions. On both land and sea, great distances had to be traveled. The Sitka, Hoonah, and Klukwan villagers had to travel about three hundred miles to reach the Haida and Tsimshian, and in the fur-trading days they even undertook voyages of a thousand miles to the Victoria and Puget Sound trading posts....[Coastal] Trading expeditions generally took place in May, June, and July when strong westerly winds blew constantly [Whereas interior trade expeditions were typically made in fall or winter between freeze-up and thaw, and after the fur-harvest].

Herring oil and dried herring eggs were among the most prestigious trade items, especially in the interior where they could not be otherwise had, along with dried salmon, dried seaweed, eulachon oil, seal oil, spruce root baskets, cedar boxes, medicinal herbs, fungus (for red paint), native tobacco (*ganch*, sometimes mixed with clamshells), dried clams, smoked cockles, abalone, dentalium, and other manufactures (see Thornton 2004:92; McClellan 1975). Today, many of these items are no longer commonly traded, although herring eggs, dried salmon, dried seaweed, and eulachon and seal oil are still frequently gifted and exchanged.

Trade partnerships have persisted to this day, but typically at the scale of the individual or household. For example, Roby Littlefield states,

> "I have a trading partner in Angoon. He will pay the freight to send it on a small plane to go to Angoon and when he has something special he'll send it to me. One year he sent me muktuk [cured whale blubber] (laughter). It was actually good quality. I was happy. But it was respect. It wasn't much, but it was a token out of respect." (RL)

Kin-based barter often does not involve money. Heather Powell remarks:

> "So, there's always been a trade and a barter, and it never was a monetary thing. So, when we were harvesting herring eggs, we would send some to my Grandma Emma, Emma Demmert in Angoon; she's Deisheetaan. George Davis's brother who is married to my Grandma Eva Davis, who was Chookanshaa from here. We would send them herring eggs and then eventually, her grandson came over and we taught him how to set so he could bring her food. And for us it was so important because every year they'd send us gumboots. Every year they would always remember us and think about those things." (HP)

Harvey Kitka explains:

> "Trade or barter is totally different [than selling]. You know ...like when I send [herring] eggs to Bert [in Yakutat]; he gives me some king salmon you know, but that's just kind of like trade. I send some down to Mike Douville [in Craig] and he gives me...probably almost two quarts of eulachon oil a year and that's not cheap, eulachon oil. Both of them know how much work goes into [herring] and how much time you spend trying to

> get the eggs. This is kind of almost like trade and barter, just a give and take. I sent some to Harold [in Juneau] one year and I never really asked for anything from those guys, but Harold sent $100, and he said, he knows how much work and how much fuel. ... But I've never asked anybody for money because I never sold it." (HK)

Contemporary trade partnerships may be mediated by money, especially when the recipient is of no (or distant) relation to the harvester and has little of value to offer besides money. This is often the case with urban dwellers, who may be wage-earners rather than participants in the subsistence economy. For this reason, customary trade for cash has been a reality for at least a century, and the idea of reciprocal payment (in cash or other goods or services) for at least several generations has been a practice in Tlingit culture, where surpluses generated in one household and/or community were exchanged for those in another. Even in these money-mediated transactions, however, there is usually some kind of relationship or reference that provides a basis of trust for the transaction (cf. Schroeder and Kookesh 1990). As some harvesters put it, "I won't trade with someone I don't know" (JD) or "There has to be a connection." Otherwise, there's no trust, sense of reciprocity or benefit flow, but rather "it's more like a business...And that's not subsistence." From this perspective, money is never the only medium or basis for exchange in a subsistence economy of barter and trade, even trade for cash. The social and cultural benefits of giving are part of the relationship; financial profit is not.

However, as Roby Littlefield explains, money enables subsistence production to exist in a mixed economy. "So, when people criticize others for doing herring egg harvest for money [trade], I'm like 'you live in dreamland.' If this was a Tlingit community, they would certainly get paid" (RL). Similarly, while some would not ask directly for cash as an offset for expenses, they appreciate that others may need to do so; asking for cash to recover one's expenses is considered legitimate in the subsistence economy. As Nels Lawson, who never sought cash for offsetting his expenses, states: "I did see they nailed [issued a citation to] somebody in the Douglas area. They said he was selling herring eggs. I don't know—I kind of wonder about was he recouping some of his expenses?" (NL).

While this is considered morally acceptable, "I suppose [once] you start making money over your expenses that crosses the line" (NL). Subsistence is about cultural exchange rather than profit.

This perspective on what is considered legal customary trade for cash was shared widely. Ken Grant notes:

> "[M]y good friend George Bennett in Sitka, he'll fill up a box and send it to me, no questions asked, but I know he had to do a lot of work and all of that. So, I'll give him cash, just to cover his expenses. It's not a purchase or anything. ... And I don't always have the cash when I see him, but next time I see him if I have a little bit of cash, I'll just give it... I'll say thank you for the herring eggs. And it's just family tribal thing. But he boxes it up and takes it out to Alaska Airlines, sends it to me." (KG)

Such transactions are not considered commercial or commercializing, but rather as Ken Grant suggests: "It's just gratitude."

Negative reciprocity—is when one takes resources, perhaps under the pretense of need or to further share, or with expectation to reciprocate, but has no real means or intention to share or return the gift. At base, negative reciprocity is a kind of theft. Negative reciprocity may occur in production, distribution, and consumption. It is generally uncommon because of the negative repercussions and disrespect that befalls those who engage in this form of "free riding" or selfish manipulation of the production or distribution system.

In the production of herring eggs, negative reciprocity is characterized by those who may steal hemlock branches set by others to harvest eggs. Generally, this form of negative reciprocity is associated with those who are either getting paid for egg gathering or are believed to be gathering eggs to potentially sell—in short those who are incentivized to abuse the subsistence fishery for profit. One group identified with the first category are those that harvest eggs for the Southeast Herring Conservation Alliance (SHCA or the Alliance), a group that represents the interests of the sac roe industry, which hires egg harvesters to provide eggs for subsistence users. The Alliance does this ostensibly to ensure that everyone who wants eggs can have access to herring eggs. The Alliance's distribution is also a public means to: 1) demonstrate that the industry supports subsistence users; 2) measure more precisely (than ADF&G-STA surveys) the "amount necessary for subsistence" through careful weighing of the harvest (so allocation decisions do not overestimate the amount necessary for subsistence and potentially put restrictions on commercial harvests); and 3) show that everyone who needs herring eggs can obtain them, if not through direct harvest, barter, or trade, then through the Alliance's distribution program (thus demonstrating that there is no significant restriction on subsistence uses). To support its distribution system, the SHCA typically hires one or two boats to set branches and gather eggs. It was suggested by several interviewees that these industry-paid harvesters are incentivized to steal branches because it is an efficient way to add to their tonnage, which is how their pay is determined. As Chuck Miller observed:

> "There's like a co-op of seiners [that pays a person here in town to represent them]. That person here in town hires people to get tonnage for their—get herring egg tonnage for them to distribute to the communities to make themselves look good. At least that's the way I see it, to cover up the fact that they're overharvesting herring. It's kind of disheartening.
>
> Those people actually steal sets from us locals, which really ticks us off because there was two—there was one in particular, I'm not going to say who, or which boat it was, but there was a particular boat that came through here two years ago, got caught red-handed twice stealing people's [herring set] buoys, knowing for a fact that that wasn't theirs. They were just trying to get their tonnage so they could get paid. That's not ok. I don't like to mention anybody's names. I think they try to pull a fast one on everybody and do whatever they can to manipulate the system to make themselves look good to... It doesn't seem very honorable. It doesn't seem very ethically responsible to do that. I don't agree with it." (CM)

Another harvester who delivers eggs to Hoonah, Vernon Hill, commented:

> "The seine fleet down there through Silver Bay Seafoods was getting money from each seiner down there, up to $2,500 I guess, and I think there's 48 permits. You can add up the numbers, it's a big chunk of money. And they've got certain people to go around and put trees out and they get them to send them to the villages. And they've been weighing them at the dock down there and at the float or on the boat they're getting them with. And they're [the harvesters are] getting either a dollar a pound or two dollars a pound, I'm not sure. But there's incentive there, the more they get, the more money they're going to make ...And last year, we actually caught one of them throwing one of our buoys out of their skiff you know." (VH)
>
> "...I've never been involved in that group that need to get the money and all that. And even if they asked me to I wouldn't do it because—you know if you accept that money from the [seine] fishermen down there and say there is a total collapse of the fishery, you're going to be part of that problem too because you took the money." (VH)

Vernon Hill has a friend (NN) who is a commercial sac roe seine permit holder that has contributed to this industry-sponsored egg distribution program, but is changing his attitude having seen its impact on subsistence harvesters and Native distribution values:

> "[T]his year, I talked to him, I've been telling him about how it's been getting worse every year on the roe; getting the eggs, harder and harder. Their [SHCA's] organizational meeting is this year, and he told them that he doesn't want his money taken out of his check this year. And he said that the Native people don't want their eggs delivered to their door. ... So, I'm glad he did it. I was hoping more of the fishermen would do that too." (VH)

Others spoke of this effect, too, as well as a more general phenomenon of industry-paid harvesters "crowding out" subsistence users by laying lots of branches in the favored areas. This practice takes away opportunities from, if not the very branches of (either through theft or careless hauling), subsistence harvesters. Jeff Jackson, one of the SHCA paid harvesters, elaborates: "[T]heir [STA's] concern is that we're putting too much—we're dumping too many trees in one bay [pushing out] their traditional harvesters. And I can appreciate that."

This crowding out is exacerbated by the more general contraction of the extensity and duration of herring spawning in Sitka Sound, which necessitates harvesters crowding together in fewer choice areas at the same time. The paid branch layers can afford to get there first, however, and also to lay more branches, so this becomes a form of negative reciprocity within the subsistence system through crowding out of valuable areas and opportunities to practice the traditional activity of herring roe harvesting in favored sites.

As Wade Martin has observed over many years:

> "And you know this Herring Alliance [SHCA], you know they sit there they, these processors and these fishermen put up money to send these out-of-towners and hire locals [to] go out and put branches in the water and uh, I've seen places that were like an automatic, they spawn every year, like in Crow Pass. And they hired these guys to go out there and they totally choked up the whole place with buoys and they totally displaced the subsistence users." (WM)
>
> "They used these guys [paid egg harvesters] who put a feather in their cap at the Board of Fish level to show that, oh yeah, there's plenty eggs to go around, and can justify their existence by basically turning a subsistence fishery into a commercial entity ... hiring people to go and produce product.... But you look at what they give away and it's sickening. It's almost all tree... it's all green and they act like they do everybody a big service and they enable people. They're lying to people, but it's just all, it's

just a big game, you know, they're just playing the game to justify their existence." (WM)

"I was out there the other day, and someone was commenting on Sitka Chatters, 'oh look at all these buoys out there,' and out there in Crow Pass those guys in the Herring Alliance filled up coves everywhere in Crow Pass and it didn't even spawn there this year. They wanted to be the first ones in there and they totally achieved their goal, they totally displaced the whole area..."

"There's hundreds of buoys everywhere out there... the two most important coves to us are so choked up you couldn't even put a tree in there. They totally displaced the subsistence users. You go over here and same thing between Promisla and Eastern Bay it's all the same thing, they just totally displaced the subsistence users this week." (WM)

"They thought they were gonna be smart [and set their branches early], and I was watching them, and I said hey, you don't put these in until it spawns because they're just gonna sit here and get brown and you'll just have to cut them loose and start all over again. They were like, 'Mind your own damn business, we're not gonna listen to what you have to tell us.' Sure enough, never even spawned there." (WM)

"It's bad enough that they take the fish out of the water seining but then they send all these damn paid monkeys out there to displace us and stress us out because we have nowhere to put branches and there's no place out there during that time of the year that we have a place to turn to that we could actually call a subsistence area. They commercialized the herring egging and they're getting paid for it." (WM)

"It's just a feather in their cap. It's all they're doing. Justifying their existence. And what better way than to go right to the source and give eggs out? They're all [the seiners are each] putting up a thousand bucks, you give away $1,000 and all the sudden you get to maintain

> that quota and they're probably getting it back a hundred-fold, you know? It's a small price to pay." (WM)

In this formulation, the price that seiners pay for the industry distribution is seen as worth it to defend the status quo. But from the local Native harvesters' perspective it is a form of negative reciprocity.

Beyond the paid harvesters, there are individual harvesters who are sometimes referred to as "egg rustlers" or "egg poachers" because they intentionally (or at least that's the perception) steal others' trees, or sets, that have been laid for spawning herring. (Note that sets may be marked with buoys or not to show ownership. Some harvesters have found that not marking their sets makes it easier to conceal them, thereby protecting them from potential poachers.) Generally, this term is applied to those who sell eggs, and are thus driven by the potential to profit monetarily from maximizing their branch-to-effort ratio. Very few people are explicitly labelled as engaging in this activity, so it is likely limited, but a single rustler may impact multiple users' sets. This kind of negative reciprocity is viewed essentially as a problem of greed, which coincides with a "commercial attitude" towards harvesting eggs.

Once harvesters return to the dock, negative reciprocity may be manifested unto those who take more than their fair share of the load, if it is unregulated. Some harvesters will on principle not restrict those collecting eggs from their boatload, because they see it as disrespectful to assume that anyone would need to be regulated, that anyone would take more than they need.

In consumption, theft may mean hoarding for oneself or one's own family, rather than redistributing (or equitably dividing) eggs in secondary distribution.

Roby Littlefield speaks about the importance of bundling shipments to save on costs and also worries that distribution on the receiving end may not end up being equitable. She remarked, "What's the point in paying the freight on [shipping separate boxes]? Cause the freight is almost the same if you send half a 50-pound box or a 20-pound box or a 50-pound box—almost the same freight; you might as well take advantage of the cost [of bulk shipping]" (RL). However, this may lead to another form of negative reciprocity: selling donated or gifted goods, in this case taking shared eggs and selling them rather than sharing. As Littlefield learned, "Sometimes there's people that call me like from Anchorage and they're young people and it's like I know their grandparents maybe and I want to share, I want them to share with their community, the Native community, and I'll send them a 50-pound box but when I find out they were actually taking eggs and selling them I won't send them eggs anymore" (RL). This secondary selling seems to be a rare phenomenon, however.

Another problem that occurs is when friends or relatives say they are going to pay the shipping and then do not for one reason or another. "Eventually my husband started complaining about the expenses that I went through. Thousands of dollars of freight... sometimes" (RL). This too is a form of negative reciprocity, if there is a prior commitment from the recipient to cover shipping costs. The problem is mitigated to some extent now by a new Alaska Airlines policy that makes it easier for recipients to pay for the air shipping of their eggs directly. As Littlefield states, "Lately, Alaska Airlines has made it simpler for people to pay their own freight. They'll call Alaska Airlines and leave their credit card number and then all I do is just deliver it and I'm like, whew!" (RL).

Complicating Factors in Distribution

Mixed economy distortions

The need for money in subsistence production is a reality in the modern "mixed economy," but in terms of gifting it remains a complicating factor. Many people are uncomfortable with taking money themselves, even if they don't know the person who seeks herring eggs. Chuck Miller explains:

> "People always kind of throw me for a loop when they say, 'Oh, I heard you had herring eggs.' 'Oh, yeah. Yeah, I actually got lucky. Been a good day.' 'Oh, would you mind if I just had a little bit. [they said]. I said, 'Well, you can have a lot if you want.' 'What? Really? How much?' Like, I get insulted. I'm like—I try not to be like too defensive. I'm like, 'Well, actually I don't ask for money. Just go down to my boat and get some.' But actually, inside

> I'm going, 'This is not right because you got people who actually sell that stuff'" (CM).

Obviously, the principle of generalized reciprocity can only work in a community that has ongoing relations (typically face-to-face). But the relationship between generalized reciprocity and barter or more balanced reciprocity is increasingly one of degree rather than kind. And, as anthropologists have long maintained, sharing, gifting, and other forms of exchange are ultimately a form of communication, a vehicle through which to send messages. Some of the most important messages from a social capital point of view are: "We're thinking of you," "We love you," "Thank you for ____," "Stay [or be] in touch," "I respect you," "Remember why we love herring in Sitka," etc. Heather Powell elaborates:

> "For many, many years, we would, just one day, get a phone call and it would he Harris Air or Seaplanes, or LAB, back in the day. So, we would get a call you know, and they would tell us ...: 'You have a box here,' and we go pick it up and it would be from Edith Bean's granddaughter, Val Vealer and Bill Vealer. And Bill is Betty Houston's grandson, you know, Herman Davis's sister. Call, and got a box, open it up and 50-pound box of king crab legs. And he'd say, 'We love you. We're just thinking about you. Can you go over and share with Grandpa?' It would be my job to shuck it and make a king crab salad. And then everybody would come over, you know. And as everybody came over we'd shuck it all, and we'd put it in freezer bags and we'd share it throughout the year, you know.
>
> So, what does that mean? That means that they think of you. That means that they love you enough to share.... So, in return in March, we send a box of eggs over. We pay the freight, everything the same they did for us. And you never think twice about it because we never had Christmases and Thanksgivings in our culture you know, that wasn't until contact...but we knew we a loved each other. So, we were able to do those things and share those things." (HP)

Mere economic transactional approaches to acquiring herring eggs, including those involving trade for cash, still provoke a certain amount of discomfort, as referred to by Chuck Miller, Roby Littlefield, and others in the above discussion on generalized and balanced reciprocity. This in part is because the messages of gifting and sharing, as emphasized by Heather Powell, may become distorted.

Sharing and Reciprocity versus Charity Distortions

While the vast majority (>80 percent) of production and distribution of herring eggs occurs among individuals, households, or kin groups, based on the principles of generalized or balanced reciprocity (with tolerated negative reciprocity as well), the issues of sharing and exchange are complicated by the rise of organizational donors. These organizations, including Sitka Tribe of Alaska, the Southeast Herring Conservation Alliance, Hoonah Indian Association, Sealaska, and the like, are all relatively new players in the distribution of herring eggs.

If they operate as charities without regard for an individual's ability to harvest eggs, then there is the possibility that they may "out compete" individuals or kin groups. The latter may not have enough funds or capacity to give eggs away, but rather may need to recoup some costs for shipping or fuel through exchange (including trade for cash). In these circumstances the charity organization can undermine more customary and traditional subsistence distribution and exchange mechanisms. If the charity approach is very successful in "delivering eggs to the dock," then economic logic suggests that subsistence harvesters eventually might find it more favorable, from a risk and expense point of view, to get their eggs for free from the SHCA program rather than face the risks and opportunity costs of going out themselves. This, in turn, could lower participation and enskilment of Native harvesters into the egg fishery and distribution system and lead to a decline in harvesting and processing skills as well as the ethics and ethos of distribution.

For this reason, as noted above, the industry egg-gathering program, which distributes eggs without regard for need or ability to harvest, is seen as "crowding out" subsistence values which emphasize developing skills to gather for oneself and to share, barter, or trade according to Native values of generosity, just distribution (e.g., to those who cannot harvest themselves) and exchange (e.g. barter for items of comparable value from other places of family or interest).

Another factor countering the triumph of economic efficiency logic when it comes to egg distribution, as the survey of harvesters shows (Chapter 4), is the fact that harvesters prefer their own eggs because they are prepared and given with care and according to cultural principles. Harvesters generally have a low regard for the SHCA industry-sponsored program for two reasons. First, the eggs are often of inferior quality, perhaps because they are processed as paid work for the purpose of generating quantity rather than the quality embedded in individual gifting and reciprocity. In other words, there is little sense of the recipient in mind. Second, as suggested above, many traditional harvesters feel the industry undermines the values of the subsistence culture by allowing able-bodied persons of means, who could harvest their own eggs, to get them for free. They prefer organizational donors that prioritize recipients, as the Sitka Tribe does, by giving first to those unable to harvest due to age, health, or means, and also put aside eggs for donation to community events. While most tribal citizens are reluctant to publicly criticize a charitable organization that is giving away eggs, they nevertheless tend to prefer their own cultural models of distribution, and if the two models clash, the charity model may crowd out high value subsistence skill building and socioeconomic network formation.

Management Distortions: What is the Amount Necessary for Subsistence (ANS)?

The ADF&G and Alaska Board of Fisheries have advanced a concept called the amount necessary for subsistence (ANS), to be referenced if allocations become necessary between commercial, recreational, and subsistence uses. It is designed to be a simple index showing whether subsistence users can "reasonably" meet their needs. In theory, if harvesters are not able to harvest the ANS, it is not because of lack of supply, but lack of demand, or lack of participation. In general, participation rates in the Sitka Sound sac roe fishery have been falling since 2002, in part because eggs are not present or accessible in sufficient quality and quantity in as many areas as in previous years (Thornton, et al 2010a), but also because the cost of accessing them, mainly by gas-powered boat, has increased

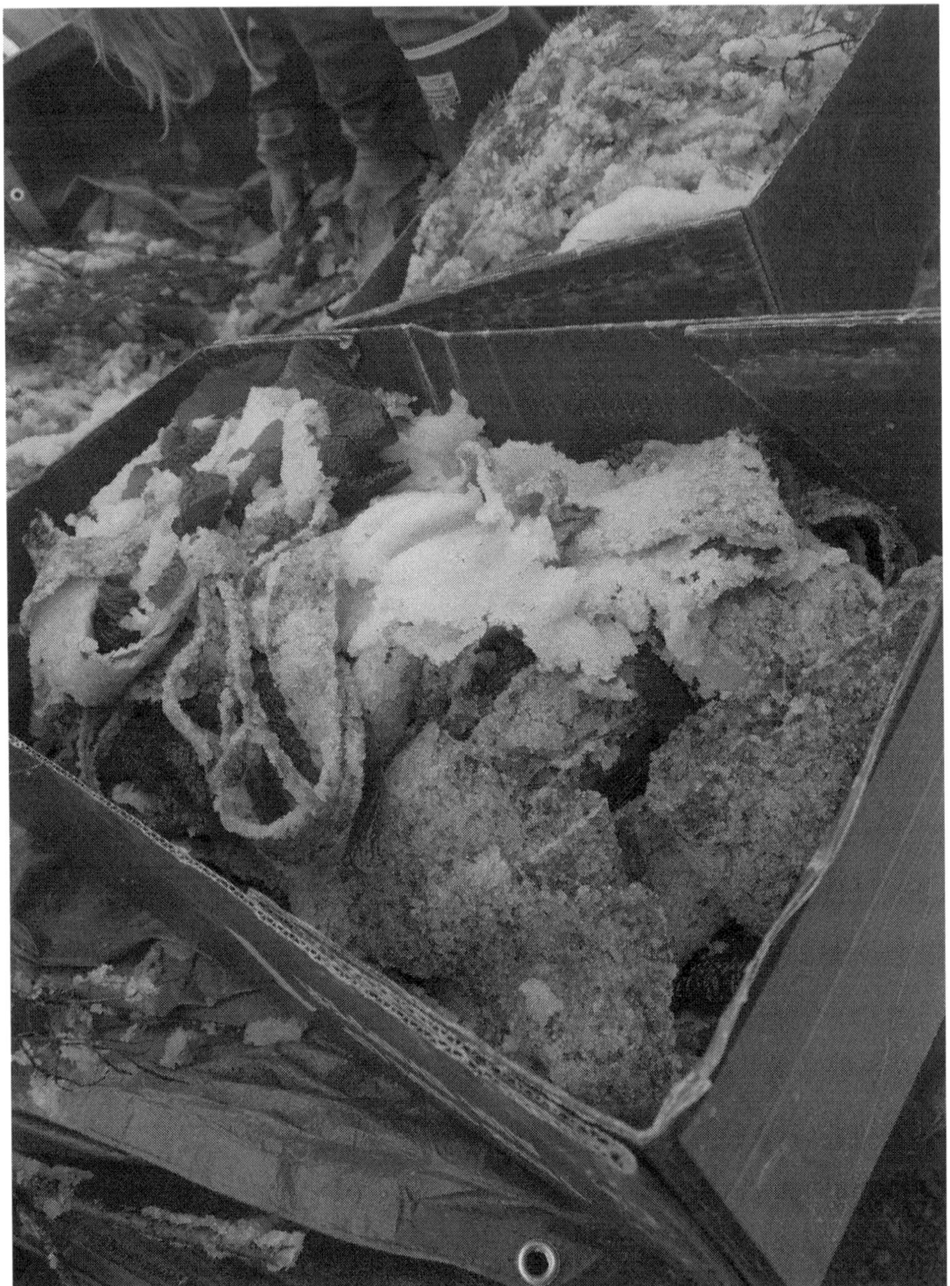

A 50 pound box of spawn-on-kelp and one of spawn-on-branches being prepared for distribution. (Photo: Bethany Sonsini Goodrich, Sustainable Southeast Partnership)

significantly. In short, there are many other factors besides simple participation and demand that are affecting the harvest from year to year.

However, the Alaska Board of Fisheries and the industry interest group tend to see reduced participation and failure to meet the ANS as evidence that there is no subsistence crisis, or even that the ANS could be revised downward, freeing up more biomass for the commercial industry. As Mike Miller attests,

> "[T]heir argument was that the Natives always overestimate [their subsistence needs] so we need to lower that amount necessary for subsistence, so it's easier to cut it, right? So, if it's a low number, we're always going to come in meeting the supposed needs of the community and then there's no good legal argument in the State [framing] to complain, you know. So, that was their effort early on." (MM)

The Sitka Tribe favors the highest harvest numbers as the base ANS because they have been unable to meet their demand for herring eggs, except in the years when the ANS was met or exceeded. For them, it is a crisis when they cannot satisfy both the local demand and the customary and traditional distribution to other communities. However, there is little pressure to raise the limit, as the formula for determining ANS is based on limited data from surveys completed in 2002 (Sill and Cunningham 2017). In addition, most Natives interviewed think the subsistence priority should be a temporal and spatial one, rather than a harvest minimum, because needs change with fluctuations in population, accessibility, and so on, as does demand (such as numbers of ceremonies, etc.). To safely accommodate all of these factors, or just to reflect historical customary and traditional levels of harvest and distribution, the ANS would need to be much higher.

Customary and traditional harvests had no pound limit, minimums or maximums, but rather were regulated by need in relation to supply. As Nels Lawson states: "I don't think it [an ANS] should be set...what we should do is monitor ourselves like the Uncles used to do. When it comes to the herring fishery, I'm always leery about the state imposing further restrictions on the fishery. ... I think the Tribe would do a better job" (NL).

Although, ostensibly, the ANS is not a limit but a minimum, it can function as a restriction when commercial harvests are allowed to rise according to a theoretical biomass that does not materialize. Raising the amount necessary for subsistence to a theoretical maximum based on tribal and other subsistence user needs would be a more conservative way to manage the fishery. Unlike commercial fishers, subsistence egg harvesters are not trying to maximize their quota, but rather are simply trying to get enough to enjoy and distribute in a way that enhances social and cultural benefits. Subsistence harvesters feel that this is an inherent socio-cultural value, which should never be at risk of being restricted unless there is a species conservation imperative. TThis kind of Maximum Sustainable Benefit (MSB) model would seem to be closer to the intent of ANILCA, the Federal subsistence law, and to Alaska's constitutional provisions for renewable resources than the current MSY model for commercial sac roe harvesting.

CHAPTER 4: CURRENT PATTERNS OF EGG DISTRIBUTION

"We depend on Sitka now. Everyone does."

—Doug Chilton, Juneau

Photo: Bethany Sonsini Goodrich, Sustainable Southeast Partnership.

Current patterns of distribution were investigated both qualitatively, through a review of the literature, interviews and profiles, and participant observation, and quantitatively through data generated by the ADF&G-STA harvest survey, administered since 2002 (with selected analogous harvest surveys carried out earlier by ADF&G; see Schroeder and Kookesh 1990 and Appendix 1).

Systematic data on distribution beyond statistics concerning giving and receiving of herring eggs by individual households have been limited to supplemental questions, introduced to the ADF&G-STA survey since 2010. In 2015-16 and again in 2018, the investigator began working with ADF&G to gain further information on the extensity and types of distribution arising from the Sitka Sound sac roe fishery. In early 2018, a supplemental questionnaire was designed for this study, to be appended to the annual ADF&G subsistence herring harvest survey for 2018 in order to gain further perspective on distribution patterns. This supplemental survey can be found in Appendix 2 of this report. The 2018 supplemental survey was administered to 27 participants identified from a harvesters list maintained by Sitka Tribe of Alaska in conjunction with the ADF&G, Division of Subsistence. The supplemental survey was not taken by harvesters who reported no harvest in 2018 or who chose not participate in the survey or supplemental survey for whatever reason.

Thus, there are several limitations to the survey. Firstly, the sample of 27 Sitka harvesters (representing 27 households in Sitka) is a fraction (perhaps one quarter) of the total active harvesters, which is approximately 100 according to ADF&G-STA records. Secondly, it was also a poor year for subsistence harvesting of herring eggs, in general, which translates not only into fewer successful harvesters, but also into lower participation rates and levels of distribution. Thus, the 2018 harvest year is very likely not representative of typical harvest years and distribution patterns. Thirdly, another limitation of the survey is that it targets egg gatherers rather than recipients, although the latter were sought out separately for semi-structured interviews as part of this study. Interpretation of the results must therefore be balanced against these limitations, which are mitigated to the extent possible by reference to related data from previous surveys, interviews, and the broader ethnographic literature.

One relevant question that has been asked consistently about the nature of distribution from 2010-2018 concerns the range of communities that receive herring eggs from Sitka Sound harvesters and how much. Raw data for this survey were provided by Lauren Sill (pers. comm 2019), and the results have been compiled and summarized in the Overview Table below, with the last column describing the overall trend of distribution in the community. Harvesters were asked where they distribute herring eggs, to answer by specifying communities and, if they chose, amounts sent (# of boxes). [1]

To get a sense of amounts distributed, a simple summation was used to register if distribution to a particular community was Low (< 5 participants or boxes), Medium (> 5 and < 25 boxes) or High (> 25 boxes). In addition, the "Highest" distribution-receiving community was identified each year, which was consistently Sitka, except for 2017 when a large number of boxes of eggs (101 in total) were recorded as being sent to Angoon, while only one respondent specified how many boxes s/he distributed in Sitka. The last column of the table identifies the distribution trend over the eight years. Blank fields indicate that no data were recorded in the original survey.

Of the 41 communities recorded as receiving herring eggs from Sitka Sound harvesters, it might be expected that Sitka is consistently the highest, with neighboring village communities such as Angoon, Hoonah, and

1 Data on amounts was not consistently recorded, however, and is missing altogether for some communities. Inconsistent sampling for the survey also makes quantification of the results less reliable. Thus, for this study, the data have been compiled to compare distribution trends only, based on a composite of how many of sampled harvesters sent eggs to a particular community and how much (where listed) was sent (i.e., number of boxes).

Overview Table: Extensity and Volume Trends of Sitka Sound Herring Egg Harvester Distributions 2010-2018

Community	2010	2011	2012	2013	2014	2015	2016	2017	2018	Trend
Anchorage	High	High	Medium	High	High	Medium	Medium	Low	Low	High-Medium
Angoon	Medium	Medium	Medium	Low	High	Low	Low	Highest	Low	Medium
Barrow		Low			Low	Low	Low			Low
Bethel	Low	Low				Medium	Low		Low	Low
Craig	Medium		Medium		Low	Low	Low			Medium-Low
Fairbanks		Low							Low	Low
Dillingham					Low	Low	Low			Low
Haines		Low	Low		Low	Low	Low			Low
Homer								Medium	Low	Low
Hoonah	Medium	Medium	High	Low	High	Medium	Low	Low		Medium-High
Hydaburg	Medium	Medium	Low	Low	Low	Low	Low			Medium-Low
Juneau	High	High	High	High	High	High	Medium	Medium	Low	High
Kake	Medium	High	Medium	Medium	High	Low	Low	Low		Medium-Low
Kasaan					Low					Low
Kenai	Low									Low
Ketchikan	High	High	Medium	Medium	Low	Low		Low		High-Low
King Cove						Low				Low
Klawock	Medium	Medium	Medium	Medium	Low	Low				Medium-Low
Kotlik							Low			Low
Kotzebue							Low			Low
Metlakatla	Medium	Medium	Medium	Medium	High	Low	Low			Medium
Mountain Village									Low	Low
Nome		Medium			Low				Low	Low
Petersburg	Low	Low				Low	Low	Low	Low	Low
Point Hope		Low								Low
Ruby		Low								Low
Saint Mary's						Low	Low			Low
Savoonga					Low					Low
Saxman						Low				Low
Sitka	Highest	Highest	Highest	Highest	Highest	Highest	Highest	High*	Highest	Highest
Tuntutuliak						Low	Low			Low
Unalakska						Low	Low			Low
Wrangell							Low			Low
Yakutat	Low	Low	Low	Medium	Low	Low	Low	Low		Low
Washington	High	Medium		Medium	Medium	Medium	Low			Medium
California					Low	Low	Low			Low
Oregon						Low				Low
Montana						Low				Low
Florida					Low					Low
Whitehorse						Low				Low

Kake showing strong distribution ties. Also showing strong distributions are regional centers such as Juneau, Ketchikan, and Anchorage, and Washington State also ranks medium to high in receipt of herring eggs from Sitka Sound. Perhaps one surprise that emerges in the data is the strength of Metlakatla's ties to the Sitka Sound herring fishery through distribution, a longstanding trend that we corroborated through interviews carried out in Metlakatla.

The remaining data summarized here are results from the 2018 supplemental survey and are organized by question number. The first set of questions concerns from where respondents (derived from the ADF&G harvester list) receive eggs other than their own harvest.

Table 1. From where have you received eggs other than your own harvest? (n=27)

Source	# of responses	%
Relatives	2	7%
Friends	6	22%
Sitka Tribe of Alaska	11	41%
Sealaska	0	0%
Southeast Herring Conservation Alliance	5	19%
Other	1	4%
None	11	41%

Table 1a. What are your preferred sources? (n=27)

Source	# of responses	%
Relatives	0	0%
Friends	0	0%
Sitka Tribe of Alaska	0	0%
Sealaska	0	0%
Southeast Herring Conservation Alliance	0	0%
Other	0	0%
Own harvest	26	96%
Whoever has the thickest eggs	1	4%

Tables 1 and 1a show that even harvesters of herring eggs may receive herring eggs from other sources, and which sources they prefer. Overwhelmingly harvesters prefer their own harvested eggs (96 percent) over others, the exception being if someone else managed to procure particularly thick eggs (4 percent).

Table 2. Do you share herring eggs? (n=27)	**# of responses**	**%**
Yes	27	100%
No	0	0%

Table 2a. What is your relationship to the people you share with? (n=27)

Relationship	**# of responses**	**%**
Family	23	85%
Friends	19	70%
Community event / community members	3	11%
Elders	3	11%
Co-workers	1	4%
Acquaintances	2	7%

Tables 2 and 2a ask if the harvester shares herring eggs, and, if so (2a), what is the relationship that characterizes with whom the harvester shares? The percentage of individuals who share eggs is higher than the 80-90 percent figure for households giving or exchanging eggs. This is likely because we did not specify giving only to other households, but rather included all sharing by individual harvesters, including within their own households. This finding is supported by interview statements emphasizing that harvesters rarely consume their eggs alone or exclusively.

On the contrary, eggs are shared both within the household and widely outside the household. As noted above, the sharing, giving, and exchange figures for herring are extraordinarily high due to the iconic and precious nature of the resource, namely its exclusivity in space (almost exclusively a product of Sitka Sound) and time (present only a few days in spring when other valued wild foods are generally not available).

Responses to this question include overlapping categories (family/friends) because kinship can be broadly and strategically articulated in Tlingit and other Native communities (as typified by the joke, recited by Kiks.ádi Tlingit leader John Duncan and others, that "when it's herring time, everybody's your uncle or your auntie"). Significantly, family commands 85 percent of the sharing, but non-familial community events, Elders, co-workers, and acquaintances also receive significant levels of distributed herring eggs.

This question was important in showing both the extensity and intensity of giving in areas around Alaska and beyond. One pattern that is evident is the importance of Alaska Native regional centers, such as Juneau, Anchorage, Bethel, Kotzebue, and Barrow as hubs for distribution. (Sitka also can be classified as a "center" in that it is the locus of the

Table 2b. In what communities do they live? (n=27) *(regional centers in bold)

Southeast	# of responses	%	Other	# of responses	%	Outside	# of responses
Juneau	13	48%	**Anchorage**	12	44 %	Seattle/ WA State	3
Metlakatla	6	22%	**Fairbanks**	5	19 %	California	2
Kake	3	11%				Whitehorse	1
Haines	3	11%	YUKON/KUSKO				
Wrangell	2	7%	Mountain Village	1	4%	SOUTHWEST	
Ketchikan	2	7%	Saint Mary's	1	4%	Dillingham	1
Petersburg	2	7%	Kongiganak	1	4%	Dutch Harbor	1
Angoon	2	7%	Quinhagak	1	4%	Sand Point	1
Hoonah	2	7%	Tuntuntuliak	1	4%		
Yakutat	2	7%	**Bethel**	5	19 %		
Hydaburg	2	7%					
Sitka	13	48%	ARCTIC				
Kasaan	1	4%	**Barrow**	4	15 %		
Prince of Wales Island	1	4%	**Kotzebue**	2	7%		
Craig	1	4%	Gambell	1	4%		
			Nome	1	4%		

harvest.) Often families that migrate out of villages move to regional centers, especially those with cultural roots (historic Native settlements), like Juneau, Barrow, and Kotzebue. These centers, in turn, may become hubs for both primary and secondary distribution. As well, relatives living in hubs, say for employment or retirement, may continue to support the production of herring eggs through donations of fuel money, shipping costs, or other means, as noted in Chapter 1.

Table 2c. Do any of these people contribute directly to your production of herring through any of the following? (n=27)	# of responses	%
No	10	37%
Yes	16	59%
Missing	1	4%

Table 2c.1. If yes, how? (n=16)	Family	%	Friends	%	Unspecified	Notes
Labor	7	44%	6	38%		Harvest/process specified
Equipment	4	25%	3	19%		Trees, boats specified
Expenses	3	19%	8	50%	1	Shipping, gas $$ specified
Other	0	0%	0	0%		

In questions 2c and 2c.1, we asked if and how people you share with contribute to your production of herring. A majority, 59 percent, said yes, and 37 percent said no. Of those contributing (n=16), just how they contributed depended significantly on whether the recipient was a family member versus a friend or acquaintance. Among family members, the most important contribution was labor (44 percent), followed by equipment (25 percent, e.g., trees and boats), and expenses

(19 percent, e.g., fuel, shipping costs, etc.). In contrast, expenses were first among contributions by friends (50 percent), followed by labor (38 percent), and equipment (19 percent), suggesting that there may be a qualitative difference in the role that family plays (less likely to contribute cash versus labor and equipment) versus friends (more likely to contribute cash towards expenses).

Table 2d. To your knowledge, do these households further distribute the eggs you share to other households? (n=27)	# of responses	%
No	2	7%
Yes	17	63%
Unknown	7	26%
Missing	1	4%

Table 2d.1. If yes, to where? (n=17)	# of responses	%
Within their community	15	88%
Outside of their community	6	35%
Unknown	2	12%

In question 2d, we also asked harvesters who gave or exchanged eggs if they had direct knowledge of further, secondary distribution of eggs that flowed from their initial distributions. Significantly, 63 percent attested to knowledge of secondary distribution, while only 7 percent said they had no knowledge of secondary distribution, with 26 percent of respondents uncertain. This is a significant finding, consistent with interview and other data, that suggests that secondary distribution levels may be as high or almost as high as initial distribution levels (80-90 percent) in some cases.

Finally, in question 2d.1 we asked those responding "yes" to question 2d (n=17) where that secondary distribution went, either within (88 percent) or outside (35 percent) of their community or unknown (12 percent).

These results are consistent with the Overview Table, though they are also based on a limited sample.

Table 3. Do you barter herring eggs? (n=27)	# of responses	%
No	17	63%
Yes	10	37%

Table 3a. What is your relationship to the people you barter with? (n=9)	# of responses	%
Family	6	67%
Friends	8	89%
Missing	1	11%

The majority of respondents (63 percent) reported that they did not barter (engage in balanced reciprocal exchange of) herring eggs, while 37 percent reported that they did engage in barter. This result suggests that most herring eggs are simply shared or gifted according to the rules of generalized reciprocity, as opposed barter or trade for cash.

Responses to question 3a suggest that harvesters may be more likely to barter with friends (89 percent) than family (67 percent), although the sample size is small (n=9). Again, this is consistent with the interview findings presented in Chapter 3 and the distribution profiles summarized in Chapter 5.

Question 3b parallels question 2b, but the results are not quite as dramatic in terms of showing that regional centers, such as Juneau, Anchorage, and Barrow, are hubs for barter as they are for sharing herring eggs. Perhaps because it is more common among friends than family, barter, it seems, is also more generalized in its geographic extent than sharing or other giving, which tracks more closely along kinship and kin migration lines.

Question 3c shows that barter also contrasts with sharing in that the majority of bartering partners do not contribute in any way to the production of herring eggs. Instead, the more balanced or reciprocal exchanges are characterized by the recipient providing a subsistence item of comparable value to the herring eggs.

Question 3d gives a sense of the breadth of items for which Sitka egg harvesters barter. Despite a small sample, because 2018 was a poor harvest year and thus barter activity was constrained, the list of bartered items is significant, including many prized subsistence foods that are not locally

Table 3b. In what communities do they live? (n=9) *(regional centers in bold)	# of responses	%		# of responses	%
SOUTHEAST			OTHER		
Juneau	2	22%	**Anchorage**	4	44%
Sitka	3	33%	California	1	11%
Metlakatla	1	11%			
Kake	2	22%	ARCTIC		
Haines	2	22%	**Barrow**	2	22%
Wrangell	3	33%	Gambell	1	11%
Ketchikan	1	11%	Nome	1	11%
Angoon	1	11%			
Hoonah	1	11%	YUKON/KUSKO		
Yakutat	2	22%	Quinhagak	1	11%
Hydaburg	1	11%	Tuntuntuliak	1	11%
Kasaan	1	11%	St. Mary's	1	11%
Klukwan	1	11%	Rampart	1	11%

Table 3c. Do any of these people contribute directly to your production of herring through any of the following? (n=10)	# of responses	%
No	7	70%
Yes	2	20%
Missing	1	10%

available (caribou, whale steak, muktuk, baleen, eulachon and eulachon oil, and red seaweed), and thus have currency as barter items. Shipping charges were mentioned by just one respondent, but this exchange normally would fall not under barter but customary trade (for cash).

In question 3e, respondents were asked specifically about secondary distribution of bartered eggs. This parallels the more general question about secondary distribution of shared eggs, with a comparable majority result, though with slightly fewer respondents (60 percent) reporting secondary distribution.

Table 3d. What items have you received in exchange for herring eggs? (n=9)	# of responses	location of partner (if given)	%
Moose meat	2		22%
Deer	1		11%
Caribou	1		11%
Whale steak	1		11%
Muktuk	1		11%
Baleen	1		11%
Seal oil	2	Kake, Haines	22%
Black cod tips	1		11%
Smelt	1		11%
Smoked fish	2		22%
Smoked eulachon	1		11%
Eulachon	2	Klukwan, Anchorage	22%
Eulachon oil	2	Haines	22%
Dry salmon	2		22%
Halibut	1		11%
Dried halibut	1	Kake	11%
Berries	1		11%
Homemade jam	1	Kake	11%
Red seaweed	1		11%
Seaweed	2	Kake, Sitka	22%
Basketry	1	Bethel	11%
Shipping $$	1	Anchorage	11%

As table 3e.1 shows, the secondary distribution of eggs within the recipient community (83 percent) slightly exceeded that of distribution outside the community (67 percent). However, this is based on a small sample (n=6), so the differential pattern cannot be said to be significant.

Questions 4 and 4a concern prioritization of herring egg giving under varying conditions of harvest. For most harvesters (78 percent), there was a priority list of people for whom they supplied eggs regardless of the amount harvested. These were primarily family members and most often Elders (Table 4a),

Table 3e. To your knowledge, do these households further distribute the eggs you share to other households? (n=10)	# of responses	%
No	1	10%
Yes	6	60%
Unknown	2	20%
Missing	1	10%

Table 3e.1 If yes, to where? (n=6)	# of responses	%
Within their community	5	83%
Outside of their community	4	67%
Unknown	0	0%

Table 4. Are there special people who you always give eggs to regardless of the amount harvested? (n=27)	# of responses	%
No	5	19%
Yes	21	78%
Missing	1	4%

who would likely receive eggs if there was any surplus available, although quantities would vary according to the level of the surplus. Interview data suggest that those family members too young or disabled to harvest and would-be harvesters living outside the community (thus lacking access) were also frequently prioritized, while those otherwise capable of harvesting (i.e., having the skills and means to access herring eggs) were not prioritized for giving.

Question 5 establishes that, beyond those recipients for whom giving of eggs is prioritized more or less regardless of the surplus harvested, the amount of eggs distributed varies in lower harvest years versus high harvest years for 78 percent of Sitka Sound herring egg harvesters. The following are some of the stated reasons for how and why distribution varies in higher versus lower harvest years. These statements are taken

Table 4a. If yes, how are they related to you? (n=21)	# of responses	%
Family	2	10%
Grandpas	2	10%
Wife's grandma	1	5%
Aunts	3	14%
Elder aunt	1	5%
Uncles	2	10%
Elder uncle	1	5%
Parents	4	19%
Mother	4	19%
Sister	1	5%
Brother	1	5%
Children	1	5%
Son	1	5%
Cousins	1	5%
In-laws	2	10%
Friends	3	14%
Elders (Tlingit relations)	1	5%
Elders	4	19%
Disabled neighbor	1	5%

Table 5. Does your distribution of herring eggs change in a higher harvest year vs a lower harvest year? (n=27)	# of responses	%
No	4	15%
Yes	21	78%
Missing	2	7%

directly from the interviewers' notes, and thus are telegraphic, often in the first person, and not always complete or unambiguous. Bracketed notes are meant to clarify responses to the extent possible.

- Don't have enough freezer [space], don't have enough [eggs]. Run out.
- [Gives to Elders/people who can't afford.] Amount that [other people require is?] not same. [So] some people, they get less.
- Don't send to other people [in a low harvest year].
- Haven't seen a high harvest in decades. When cooking, invite people to eat eggs [rather than giving away].
- Less [shared/given/bartered/traded] when less harvest.
- Able to reach more enrolled tribal citizens [in good harvest years].
- Still share, but a little less when less harvested.
- Fewer eggs to same people [in a lower harvest year].
- Send out way less. Used to send 70/80 boxes. Last several years 2 boxes. This year zero. [2018 being a very poor year for most subsistence herring egg gatherers]
- The size of boxes. The amount per person. More boxes if better year. People get upset if they find someone else got eggs, so don't send at all on low years.
- Less than this year, didn't share any. Don't share as much [in lower harvest years].
- It might. This is all the more they share.
- Used to get a lot and share it all. This year I decided not to eat any [i.e., shared 100 percent].
- I'll share more. This year shared more than harvest (from stash).
- It would. Maxed friends list with this year's harvest. Would have to get more friends.
- Keep going out until people say they have enough. Haven't had enough to share [this year, a low harvest year].
- Usually [ship to] Southeast [only, if a lower harvest year]. A good year—all over Alaska.
- Didn't give away as much as usual. Usually bring in 2,000 pounds. People don't distribute in low years to people who aren't close relatives. Hard to tell people this year [I'm not giving because] there was none.
- The eggs can go farther (to other communities) [in a higher harvest year].

- More to family and local neighbors - community functions [in lower harvest year, and less to those more distant socially and geographically].

While not systematic, these responses reveal a number of factors that affect the level of distribution from individual harvesters' perspectives. Overall, it seems that both the level of giving and the patterns of distribution are affected when herring egg harvests in Sitka Sound are lower than expected. Not surprisingly, people give fewer eggs if they have harvested fewer eggs, but some will try to maintain their extensive networks of distribution so everyone enjoys "at least a taste." In contrast, others will contract their distribution and ship eggs only along certain key social networks (e.g., kin) or to certain communities (e.g., local or regional distribution but not statewide) as a means of coping with "distribution stress" caused by not enough eggs being available to harvest.

CHAPTER 5: PROFILES OF HERRING EGG DISTRIBUTION

"We got a line of about say 50-60 people and they're walking off with anywhere from 20 pounds to 100 pounds [of herring eggs] apiece. We just gave them away until they were gone."

—Charles Skultka, Sitka

Wade Martin is a consistent high harvester of herring eggs and distributes his eggs widely. He formerly collected and distributed eggs on behalf of STA when he held the role of traditional foods coordinator. (Photo: Bethany Sonsini Goodrich, Sustainable Southeast Partnership)

While quantitative data tells us how much harvesters gather, give, and receive, it says little about the "how" and "why" of distribution. Thus, we include profiles of individual households and donor organizations to complete the picture of Sitka Sound herring distribution.

Schroeder and Kookesh (1990:27-35) engaged a small number of Sitka high harvesters to compile detailed qualitative profiles of their distribution and exchange activities in the 1989 Sitka Sound herring egg harvest. They found evidence of sharing, reciprocity, trade for cash, and other forms of exchange characterized here. The five harvesters they interviewed ranged in production from 1,000-5,600 pounds of eggs, most of which was given away or exchanged to villages all over Alaska, but mainly neighboring villages and urban centers in Southeast Alaska. Schroeder and Kookesh also collected data on freight shipments of herring eggs for 1989 based on airline records provided by Alaska Airlines and Bellair, as well as boats, including the Alaska state ferries. They estimated that approximately 50,000 pounds of eggs, primarily eggs on hemlock but also roe on macrocystis and hair kelp, were shipped by these means of transport, with the majority by air, especially via Alaska Airlines.

In this chapter, a series of detailed contemporary distribution profiles are compiled for comparison. However, in this case, the profiles are not limited to high harvesters, but rather also include donor and recipient organizations, such as the Sitka Tribe of Alaska, the industry-sponsored group Southeast Herring Conservation Alliance, the Salvation Army, and others that engage in primary and/or secondary distribution of eggs.

Although we were unable to get freight shipping records for this survey, as in the previous study, the airlines remain the major carriers of eggs to other communities due to time sensitivity involved in shipping fresh eggs out of town. To complement the earlier study, we focus here on how, for whom, and why eggs are harvested and distributed in both the contemporary and recent historical period.

Individual Household Distribution Profile: Harvey Kitka, Sitka

Harvey Kitka, a Tlingit Kaagwaantaan clan Elder in his late-70s, typically sets three to six hemlock trees in two or more locations in Sitka Sound each spring, depending on how the herring spawn evolves. The investigator has observed Harvey set and pull hemlock trees in Sitka Sound on a number of occasions. Although he knew 2018 was proving to be a poor year, Harvey instructed his nephew, Jerrod Galanin, to set trees at a promising spot in Promisla Bay that showed signs of spawning activity. When we pulled the two hemlocks several days later, one had pretty good spawn, the second not much. Harvey trimmed both and towed the second one a little distance to a spot he deemed to be better. The trees yielded a large cooler full of eggs, a modest return, with little thick spawn due to the poor spawning year, but better than many harvesters had done. At the dock, Harvey further trimmed up the branches to optimize for thickness and then transported the cooler back to his house. At the house, the eggs were stored outdoors in a cool, shaded place.

Distribution of these eggs was carried out with the author engaging in participant observation as follows on Sunday, April 1, 2018 (Easter Sunday).

- Nephew: Harvey phones his nephew Jerrod, who is told he can take as many eggs as he needs whenever he wants to pick them up. This is in recognition of his contribution to production (laying the trees), which was integral to the successful harvest. It is also perhaps recognition that Jerrod is mature enough to judge for himself how much he needs for his own distribution network.
- Female relative of Harvey's wife, Gina: This relative was a student at Mount Edgecumbe High School, an Alaska Native

Harvey Kitka, a Kaagwaantaan Tlingit Elder, chairs the Sitka Tribe Herring Committee and has been harvesting herring eggs since he was a boy. He is careful in selecting and placing his hemlock branches to attract the herring. (Photo: Bethany Sonsini Goodrich, Sustainable Southeast Partnership)

boarding school on Japonski Island, and previously had received herring eggs on behalf of her mother and grandmother, who live in Mountain Village in interior Alaska. At that time, they had expressed an interest in trying herring eggs and really liked them, so the Kitkas gave them a box to take back with them to Mountain Village. They appreciated this gift. Gina knew the student would be going home for vacation later in the spring, so she invited the student over to pick out some eggs, trim them further, and bag them for freezing to pick up when she returned home to Mountain Village. Importantly, the student was not just given eggs but shown how to carry out the specific processing techniques. Afterwards, the student was able to visit and have dinner with the family. It is significant that this sharing also involved "enskilment" (Ingold 2000)—the training of the young relative on how to select, trim, and process eggs to "put up" (freeze/preserve) for later use. This is part of the learning involved in herring harvest and distribution. When she returns to Mountain Village, the student can be proud that she helped produce the eggs that her family will consume. The quantity was not large, and when I asked the student how the eggs would be shared, she stated that they would be shared mainly within her immediate and extended family, and typically consumed communally in one meal or two as a special treat.

- Aunt of Harvey's wife, Gina: A gallon bag of eggs was delivered to Gina's aunt, Rene Ausbach, and family with a loaf of homemade Easter bread.
- A bag of eggs and loaf of Easter bread is laid aside for Harvey's maternal uncle, Charlie Daniels, who is nearly 89 years old (b. 1930).
- A bag of eggs is put aside to be shipped to Harvey's older brother, Herman Jr., who lives in Anchorage.
- Chuck Miller is given a bag, along with Easter bread. Chuck is a L'uknax̲.ádi clan leader who married a relative of Harvey's. He is normally a significant harvester of herring eggs, but this year has experienced poor returns. This is an example of one herring egg harvester's gifting to another egg harvester, which is not uncommon.
- Others are given Easter bread loaves but not herring eggs, as this year the supply of eggs in limited.

Normally Harvey will also send eggs to relatives and friends in Juneau (e.g., son's partner and family, Harold Martin), Craig/Klawock (Mike Douville, a trade partner), and Yakutat (Bert Adams, a Tlingit friend who served with Harvey on the Regional Subsistence Advisory Council). This year, however, due to the minimal harvest, he is sending just one box to Juneau with a relative and does not feel he has enough to send to friends in Craig/Klawock and Yakutat. His distribution of eggs has thus been limited by the poor 2018 spawning in Sitka Sound.

Individual Household Distribution Profile: Chuck Miller, Sitka

Chuck Miller is a middle-aged Tlingit leader of the L'uknax̱.ádi clan and has been an active harvester and distributor for most of his adult life. When he was growing up in Sitka, his family would receive herring eggs from his grandparents and from his Uncle Herman:

> "I'm pretty sure it was my grandparents, my mom's parents would always bring us some. And sometimes my Uncle Herman would call up the family and he'd call everybody that lived here, say, 'Hey, there's herring eggs down at the boat. Go get some.' And that's kind of how it happened after my grandparents got too old. I don't know if they got them themselves or somebody gave it to them to give to us. I'm not sure." (CM)

His mom would then redistribute them to family members in Kenai and Soldotna and friends in Juneau:

> "[M]y mom would send out stuff to her sisters up north once in a while. Because even when I got older and I started getting my own [herring eggs], I'd bring it home and she'd always want some to send to her sisters up in Kenai and Soldotna and whoever else; and sometimes her friend in Juneau, she would ship some out." (CM)
>
> "I would make sure everybody, whoever wanted some I'd put it out there, say 'Hey, just go down to my boat,' like my uncle did. 'Go down to my boat and just grab it up. Take what you want. Put the tarp back over when you're done and put the rocks on top, so the birds don't get into it.' And that's all I ever said. 'You want it; you haul it.'" (CM)
>
> "It changed. Just because people move. The same people ask for it, but they would move. Like, my sister, she lived in Juneau for a number of years and I would

> ship it to her and her husband. And then she moved, so I ended up shipping them down to California. And then she moved again, and now she lives in Washington. So, I ship it down to her. But I haven't done it in the last couple years because I got nothing to give her." (CM)
>
> "[A] person in Kake, their family would request some and I would ship it to them. The last two years I haven't been able to. Somebody else in Juneau wanted some and I'd ship it to him. They would all pay for their boxes and the shipping costs..." (CM)
>
> "[T]here's a big white tote that I usually use. It's usually used for—commercial guys kind of have them. It's a big white tote. I don't know, it's like maybe...a rough estimate, maybe four wet boxes, I guess. And usually I pile that up pretty high. And when it was doing well, I usually went through one of those...probably two of those year, sometimes three. It just depends on who wanted what. So just a rough estimate, probably what... eight to twelve 50-pound wet boxes, give or take." (CM)

Chuck Miller prepares enough trees and makes enough sets to accommodate those who depend on his distribution:

> "I always made sure I at least set out ten trees. And these weren't enormous trees. These are just—I don't know—what is that four inches maybe? Maybe four inch in diameter at the base or less, hemlock branches, trees." (CM)

However, lately, this number of trees often has not been enough because of poor spawning, theft, and crowding out of the best areas. At Sealaska Heritage Institute's Celebration 2018 in June, while leading the Sitka Gajahéen Dancers in a performance, Chuck Miller took the unprecedented step of making a public appeal on behalf of the Sitka Natives for support on the herring issue. He noted that in recent years, for the first time, he has not been able to harvest enough herring eggs to meet his needs, and this year had not even enough to distribute, even to family who depend on his annual harvest and distribution. This appeal was met with a strong ovation from the mostly Southeast Alaska Native crowd of about 1,000 spectators filling Centennial Hall in Juneau.

Individual Distribution Profile: Mike Miller, Sitka

Mike Miller is a middle-aged Tlingit harvester of the Shangukeidí clan. As a single man without children, he distributes virtually all of his egg harvest.

He ships approximately 60 50-pound boxes of herring eggs (mostly on branches) out to relatives and friends at destinations around the state, including Juneau, Hydaburg, Ketchikan, Haines, Metlakatla, and Anchorage, and as far away as St. Mary's on the Yukon River and Utkiaviq (Barrow) on the Beaufort Sea.

> "[I] got out of high school and started working towards [marine pilot] licenses and such, I was working in shipping with [a]... Japanese shipping company... involved in buying herring, Tokyo Fisheries, from 1984 to about 2001. So, I was always here for the herring seasons every year to get eggs." (MM)
>
> "And so ever since in the [19]84-85 timeframe I would get a lot of eggs. On our mom's side of the family [Shangukeidí], a lot of extended family around... Shangukeidí, the Thunderbirds... There's only a handful of [them here] ... a lot of the true blood relatives or even opposite relatives in the clans were from Hoonah and Juneau, kind of scattered around different parts of the state. And like everybody who worked on boats, you get a lot of friends that ended up in [Sitka], just from different parts of the state, you know. So, people in Metlakatla, Hydaburg, Ketchikan, Haines, Anchorage. So ... some friends would come here [and] we'd just go out and get eggs." (MM)
>
> "And even the fleet before the [sac roe] commercial fishery, I think there was a lot more Alaska Natives, so there was quite a few friends on the boats that would

> come and we would all collectively go out and get eggs to ship, you know… either take them home on their boats or ship them out." (MM)

When harvesting with friends from Metlakatla on the [Frank] Hayward's boat, they would harvest "15-20 tons of dry weight, with eggs. And… pack them fresh, except old man Frank [Hayward], and he'd always pack his in salt. Yep, and that's the way he liked it" (MM).

When the other boats from out of town stopped coming to Sitka, Miller began to distribute eggs on his own:

> "And as they quit coming, you know, it kind of shifted to having to just go out and do my own, but it'd be my schedule with the boats too. I would just go out on my own after a while and …ship what I could, which usually ended up about 60 boxes. That's about what I could—that's all I wanted to do in branches." (MM)

A lot of the recipients were not relatives, but friends who he worked or played with:

> "And a lot of guys, you know, we'd play ball together, we'd do stuff. … I was here, and I knew people, because I worked on boats, right? So, we'd see each other all around the place. But, you know, there's a lot of guys that were similar age. I think the Haywards in Metlakatla, Edenshaws in Hydaburg, and they aren't true relations, but we got to be quite close. … So, for some of those families, I got to be—they were always nice, so we always checked in, but we always got good eggs too. So, I think that was one of the things that … people always knew—that I'd [make] the time to go out and make extra sets [to supply them]." (MM)
>
> "I found an old list, and I was looking for it later. I know it's somewhere, I can't find it, but I had in, I think '90…there was 23 people that I'd ship to…mostly Elders

> in Hoonah. ... 50s and 80s [pounds] sometimes. We used to get the big boxes. I rarely would use 25s because it's just easier to pack the big ones. Three years ago, it was down to three people that were left out of that list." (MM)

He would also regularly ship to Juneau where he had multiple aunts, along with other family and friends. He might ship as many as 30 boxes to Juneau. At times he would try to take advantage of economies of scale and ship bigger boxes to fewer recipients, who would then be expected to further distribute the eggs. However, this was not always possible due the fact there were "cliques" that did not necessarily get along, or intermediaries who did not necessarily carry out equitable distributions. This was accompanied by an amusing story about his eldest aunt who tended to take a disproportionate exaction before redistributing smaller quantities to his other aunts.

> "It was a lot cheaper to ship to one [aunt], and so I'd always send four boxes over to each of the aunts [care of the oldest one] ... And after quite a few years of that, my one aunt who just passed away, she was like, 'We're always so glad we get that bag of herring eggs every year from you,' you know. And I was like, 'Bag?' And so, the one aunt had actually been keeping all the boxes and taking a bag out of each one to give to my other aunts. So, for quite a few years, I had to ship individual boxes to each person until I found one of my cousins who was going to distribute it you know. So, I had to find a real neutral person to distribute eggs [equitably] [laughs]." (MM)

Miller's distribution range is the broadest of any individual interviewed. This is due in part to the fact that his clan is less rooted in Sitka than elsewhere, in part because he serves on boards and commissions that put him in contact with people statewide from communities as far away as the Arctic and sub-Arctic, with whom he develops sharing and barter relationships, and in part because he has worked on boats for many years with non-locals who desire herring eggs.

Individual Distribution Profile: Mike Baines, Sitka

Mike Baines is a middle-aged harvester of Tsimshian and Tlingit descent who was raised in Sitka. He is single without children and considered a consistent, high harvester of herring eggs, most of which are distributed to others. He remarks, "I've sent herring eggs all over Alaska and the Lower 48 and some to Canada. And friends and family—also traded quite a few things for the herring eggs" (MB).

Baines also has "relatives [who] get it [herring eggs] in Prince Rupert [British Columbia]," (MB). Ironically, he notes, the Prince Rupert area gets a fair amount of herring spawn, but people say they prefer the "Sitka brand" of herring eggs.

Of the 80-90 boxes of herring eggs Mike ships out of town each year, some people will "get two or three boxes" or in the case of one individual who assists him with secondary distribution, "six or seven boxes sometimes." He estimates that the secondary distributor might give about a dozen to other people or households.

Baines ships to Anchorage, Juneau, Barrow, Ketchikan, Metlakatla, Prince Rupert and Port Simpson, and Wrangell. His connections in Metlakatla and British Columbia are his Tsimshian relatives, but he also has statewide connections through his professional work, including as STA chairman. When asked to break down how much he sends where, Mike estimates, "From what I've sent off... the last few good years... I would guess 25 percent at least [goes to] Juneau and Anchorage" combined. Another 25 percent goes to Ketchikan, generally to one person, a female tribal leader, who then redistributes to the community. He remarks, "I was there for something else but went to lunch at the ANB [hall] there and they had some herring eggs that I sent to her and she [had] donated to them. ... [S]he used to be in the KIC [Ketchikan Indian Community]. She used to be on the board or council." Another 25 percent goes to Metlakatla (10-15 boxes). Smaller amounts would be sent to Prince Rupert and Port Simpson: "I'm enrolled in my mom's band, the Lax Kw'alaams." He further notes:

> "For the last few years, Alaska Airlines said our family was, by far, the biggest shipper of herring eggs."

However, with poor spawning returns in Sitka Sound in 2018, Mike "didn't even monkey around with herring eggs at all this year" (MB).

Individual Household Distribution Profile: Charlie Skultka, Sr., Sitka

Charlie Skultka, Sr., a Haida Elder originally from Hydaburg, is among the few Natives who has participated regularly in both the herring sac roe seine fishery, as captain and crew, and the subsistence egg fishery. He has regularly sent eggs to Hydaburg, Craig, Klawock, Juneau, Anchorage, and Seattle. He notes, "and some of it even made it as far as the Philippines" (CS). He both ships by air and, when he was running a boat, delivered them directly to communities around the Southeast region.

When transporting eggs for distribution by boat, Charlie Skultka notes that preparation is important to maintain the quality of the eggs over the course of the journey:

> "We had to box them up and chill them down. Like from here to Hydaburg or here to Juneau, here to Craig, whatever, when we're transporting. We didn't have RSW [Refrigerated Seawater System] back in those days, so we'd have to like hose them down every hour, keep them as fresh as we could and when we'd get to the other end of the line they'd be as good as new." (CS)

Skultka was one of the few Natives in the commercial sac roe seine fishery, so he would collect eggs after the fishery ended, and would then use his seine boat to distribute them to other communities.

> "And we got a line of about say 50-60 people and they're walking off with anywhere from 20 pounds to 100 pounds apiece. We had, say, 25 thousand pounds of eggs aboard. We just gave them away until they were gone. But then we went and we got another load and we ran them to Juneau, about the same [amount]." (CS)

He notes that other Natives in the seine fishery also did this. For example, "Joe Demmert and his family used to take eggs toward Craig and Klawock and Ketchikan and Metlakatla" (CS). And some boats with Native captains or crew still do this, especially if they are from out of town. Seine boats can transport large quantities of eggs, but most did not try to maximize their load, but rather just took enough to share with other communities, where they had relatives without a local supply of herring eggs.

As a Haida originally from Hydaburg, Skultka's patterns of distribution are different than those of Sitka Tlingits. They are more distinctly oriented towards southern Southeast Alaska, and Haida country in particular, along with the regional centers, such as Anchorage, Juneau, and Seattle.

Individual Distribution Profile: Doug Chilton, Juneau

Doug Chilton is a Tlingit artist living in Juneau. He grew up working with his family on a herring boat in Auke Bay, Indian Cove, and other parts of northern Southeast. Herring eggs have not been available in Juneau for many years because of the crash of the spawning stocks in Auke Bay.

About ten years ago, Chilton took on the work and expense of continuing a herring egg distribution from Sitka to Juneau after ANB (Alaska Native Brotherhood) Camp 2 in Juneau was no longer able to fund the operation. Initially, Doug had helped with the ANB Camp 2 distribution program, which typically involved supporting fuel expenses for someone to go over to Sitka and harvest eggs to bring back to Juneau for distribution to several hundred community members, mainly Elders and those otherwise without access to herring eggs. Andy Ebona (pers. comm. 2018), a Kiks.ádi Tlingit leader originally from Sitka, helped establish this program through Camp 2. Before they used Doug, the ANB camp, which funded the operation from pull tab revenues and donations, would pay the expenses for a Sitka egg harvester to ship several thousand pounds of herring eggs via a state ferry, often loading them on a truck (sometimes Doug Chilton's, brought over from Juneau). Chilton assisted with the harvest operations as well as the transport.

The camp was responding to a need among Juneau's 5-6,000 Natives, many with ties to herring egg communities, for fresh herring eggs that were no longer accessible in Juneau. Camp 2 (Juneau) worked with the Central Council of Tlingit and Haida Indian Tribes of Alaska (also located in Juneau), which had a list of some 30-40 Elders. They would deliver the eggs to the Elders primarily.

When the ANB program ceased around 2007, Chilton took over distributing herring eggs from 2008-2010 at his own personal expense. He estimated expenses for those two years totaled 11 to 12 thousand dollars a year:

> "I probably never ever would have thought of doing it except for the fact that I was involved with [ANB] Camp 2, and it was something they were doing when I got there. And I saw how the Elders appreciated it … because they didn't have access. So... The first time … the guy we were working with [previously] offered to work with me on it, and at first ... I was concerned about the expense of it, and then I started to think about it and [began] talking to some of the Elders. And the personal gratification of seeing them happy outweighed the expense for me. ... And then not only that, you get these people here locally that don't have access that are not Elders for whatever reason, for them to be able to have access to it, it made a big difference for them. I mean, it's a traditional food, and we don't get a lot of traditional food around here anymore." (DC)

For Chilton there was a distribution priority:

> "First, I go to the Elders … Even if I don't know them, of course with one box you can't share with everybody. I always start with the Elders that are closest to my relatives. And then like my neighbor, he's not physically able to get out and do stuff on his own, but he loves his traditional foods, so I always make sure I bring him some." (DC)

Chilton notes that many people in Juneau, who grew up when the shores around Auke Bay were full of spawn, still have a strong taste for herring eggs, but they can't get it anymore locally. "We depend on Sitka, now; everyone does" (DC).

Distribution Profile: Mike Smith, Household and STA Harvester, Sitka

Mike Smith, a middle-aged Kake-born Tlingit now living in Sitka and employed by the Sitka Tribe, illustrates that Southeast Natives and their communities are neither isolated nor static in terms of their relationships with the herring egg distribution system. A recipient of herring eggs as a child in Kake, Mike moved to Sitka in 2008 and has now become a harvester and giver of herring eggs as both the head of a household and an employee of STA as a traditional foods specialist for their Traditional Foods Program (see next profile).

Growing up in Kake, Smith remembers that his family would receive herring eggs:

> "[W]e used to get some; my dad put in orders from here [Sitka]. We'd get about six to eight [50-pound] boxes at least. … He would call up one of his friends from here and just tell him he'd pay for the boxes and freight" and would help pay for fuel.
>
> "We would pretty much eat all of it ourselves [as soon as it arrived], since we didn't have access to it, it was—my mom would cook it. We'd either have it as a lunch or she'd make a herring egg salad" (MS). Mike grew up in a large household with three brothers and two sisters and a nephew who lived with them. His parents would "always invite over their friends and they'd share with them." (MS)

For a span of thirty years, his family received herring eggs "pretty much every year" from Sitka by boat or air freight.

> "I can still remember a lot of people, you know plane after plane coming into Kake but just loaded with herring egg boxes. … "There would be a plane-load, so anywhere from like 12 up to 20 boxes I would guess."
>
> "[Y]ou could always tell when it was herring egg season, the dock would be…there'd be a stack full of boxes." (MS)

Smith estimates that for every six to eight boxes that arrived in Kake, it would be distributed to "at least another five to eight families. … Because we'd share—we have a pretty big family in Kake and we'd share with everybody. Usually when they'd have a Kake boat come over, they'd do really well, and you know, load up the hatch and pretty much feed the whole community" (MS). He estimated that distribution would be somewhere between five and fifteen thousand pounds.

Smith says his role as traditional foods specialist for the tribe also brings a sense of satisfaction and purpose. "Not everyone can get herring eggs for themselves," he notes, "Some do not have boats or equipment," or they are not able due to health or age issues. Thus, he thinks the distribution that the tribe carries out is important to make sure that those people who cannot harvest eggs for themselves do not lose their connection to this important first fish of the spring.

Institutional Harvest Profile: Sitka Tribe of Alaska Subsistence Foods Distribution Program

The Sitka Tribe was a pioneer among Southeast Native tribes in initiating a Tribal Foods Program more than 20 years ago. Previous heads of the Traditional Foods Program, such as Wade Martin, were often selected on the basis of their harvesting abilities and remain active as individuals in the harvest of herring eggs for distribution. Today Traditional Foods Specialist Mike Smith and Jeff Feldpausch, head of the tribe's Resource Protection Department, do most of the setting of branches for the tribe's herring egg harvest, and along with other staff, who assist with the processing and distribution of eggs. Eggs on branches are the tribes' main focus, although they do collect some roe on kelp (*daaw*) and hair seaweed (*né*) on occasion.

Teresa Moses, T'ak̲deintaan Tlingit clan, trims eggs on branches for packaging and distribution. (Photo: Bethany Sonsini Goodrich, Sustainable Southeast Partnership)

Feldpausch explains that the tribe prioritizes Elders and other special populations in their distribution:

> We have a traditional foods policy that directs staff on the hierarchies to how to distribute subsistence resources. That hierarchy is there in case of a limited resource and how it would be distributed. What we do with herring eggs and our sockeye salmon, those are the two subsistence harvests that come in in very large quantities in season. We don't have the capacity to process them all down and put them in a freezer some place. We don't have the capacity nor the space to hold it all. So we do in-season distributions for those particular items. We start off with our Elders, according to our policy. I think right now we have over 500 Elders that qualify as an Elder within the tribe, which is 55 and older. And as the eggs come in and are available for distribution, we start with our oldest Elder and start making phone calls to try to make direct contact with them. We limit what we distribute to a full plastic store bag [2-3 gallons]. We don't have the volume of harvest for people to come in and get as much as they want. It's what we can—the volume we go with for distribution, individual distributions is enough to try to spread that harvest out. If we didn't, we would have people backing up with trucks and wetlock boxes to ship to all their relatives and that's not what our program is for...
>
> Then, we also put eggs up for the tribe for events. I think I filled a freezer this year because we went without [eggs] last year. We ran short, didn't realize the harvest was going to be over when it ended [i.e. prematurely with poor spawning] ...we had to short some of our Elders because we only got about halfway through the list before we ran out of eggs. If it's one of those years where a lot of the other harvesters are doing good...it takes some of the pressure off of us to provide eggs to folks. So, then

> it allows us to get through our Elders list and what we do then is we start putting up vacuum packed quart-sized bags for out–of-season distribution. So that we would put them in a freezer and if somebody came in for normal distribution, which they can come in twice ... a month... We would offer them a quart bag of herring eggs. So, we would have out-of-season distributions. If there's still eggs left over then we would open it up to the rest of the general tribal citizenry to come and get eggs--the same volume that we would give the Elders. I have put a limit on it as we get towards the end of the season. If the community is saturated in eggs, I've gone as far as letting people have one wetlock box. (JF)

Ordinarily, however, quantities are kept smaller, both because there is not enough to give tribal citizens more, and because if it is opened up, there may be large secondary distributions outside of town.

> Some of the concerns we've had in the past, and I think it's probably happened, is end of the season you keep bringing in eggs, and some people will come in and start filling wetlock boxes and distributing it. That's not going to tribal citizens here. It may be going out elsewhere. But the money I have in my budget is to provide for tribal citizens. So I limit them to one box. There may have also been selling of wetlock boxes. And from my standpoint within the tribe, this is subsistence, we don't sell it. (JF)

Among tribal citizens, the priority from highest first is: "tribal Elders, folks on disability, general assistance, tribal citizens, general tribal citizens living in Sitka, and then general tribal citizens outside of Sitka" (JF). Thus, cultural identity, economic, and geographical considerations are factored into the protocol. Feldpausch estimates that, on average, 300-500 tribal citizens have received direct distributions of herring eggs, often a standard-size plastic grocery bag's worth, annually since 2010 when he began harvesting for the tribe. However, some years are worse due to poor supply or theft. He notes, for example, "2017 was horrendous. We ended up with 1,200 pounds stolen and 1,260 pounds of eggs. I need a minimum of 4,000 to 5,000 [pounds] to make it work." Competition sometimes creates problems of access and congestion in choice harvest areas. In 2017,

> "All ... the quality spawn occurred in northwestern areas, Middle, Crow, and maybe Gagarin [islands]. All in that tight little circle there. We had spread our sets out. The only sets that came in of any good quality were those off of northwest Middle, the Cookie Jar area. ... it was packed. There was probably 100-150 sets in there. And when we went back to pull, we lost six of the thirteen sets that were in there [plus one that was damaged by debris]. ... So, we got six sets and got 1,260 pounds. The other six sets either went to individuals or went to the Katlian industry boat who got caught by several people taking [stealing] eggs. So, yeah, we had a very poor year." (JF)

Competition from the industry program exacerbates the problem of harvest distribution for STA:

> "If they really want to help us out, maybe they can stop saturating these areas, the core areas, stop saturating them with sets. Because what they'll do is, they set out 100-plus sets. And in order to get your sets in, in time, they do pre-setting, kind of like we do. I think we did 36 sets total this year [2018]. I know I can't get them out in a very short period of time, so we hit high productive areas that had a high frequency of spawn. The problem is, so does the industry. So that before a single egg is laid, some of these bays are just packed with sets and you can't get in there because these guys are—if they're being paid by the pound, I'm sorry, but that's a commercial fishery. So, they're out there with the incentive to bring in as much pounds as they can so they can make money. We're trying to provide eggs to Elders and other subsistence users

> and it's not a commercial thing for us. So, we asked them to—look, the industry is arguing, 'All you have to do is go out and go where the fish are at and make your sets.' So, we said, 'Well, would you stay out of the core area and not saturate it?' I think they still did anyway." (JF)

This competition, and the general decline in available spawn, leads to a poorer return on effort, according to Feldpausch. The remedy for this is often either to withdraw from the fishery or to put in more effort—i.e. more sets—further clogging the spawning grounds and leading to even poorer returns on effort overall. At this time, this scaling up of effort by some harvesters may make less aggressive branch setters less successful and perhaps constitutes a reason why overall participation rates have been in decline since 2002 (Sill and Cunningham 2017). Those who cannot afford to increase their effort may simply choose to stop harvesting altogether.

Thus, despite their active distribution program as well as others putting more trees in the water, STA argues that tribal subsistence needs "aren't met, haven't been met since 2015." (JF)

According to STA Tribal Foods Specialist Mike Smith, STA harvesters make an effort not to compete with their own tribal citizens when they are harvesting:

> "[W]e always try to get out there and you know, scan the area, do the best we can to be the first ones to get our trees in the water. And we've been pretty good about that. And then if there are other sets, we try to set away from them, even though the quality may not be as good. But most of the time, we bring in some good eggs." (MS)

At the same time, the tribe does not want to dis-incentivize capable harvesters by "getting their eggs for them"; hence the priority distribution for Elders and those who cannot otherwise harvest. STA also limits its number of sets; in recent years it has been between 15-17 trees.

Heather Powell (2015 interview), who previously worked for the Sitka Native Education Program (SNEP), a subdivision of STA, appreciates the Tribal Foods Program's herring egg distribution system as a supplement to the traditional Tlingit protocol:

> "Traditionally [the] protocol is that every son-in-law will make sure that his wife's family has food. Every nephew will make sure that his uncle that taught him has what he needs. And as a community I think that the preferences...[ensure] that elderly folks are taken care of [as priority], but there's also the community sense of making sure we have enough for when Alaska Native Sisterhood hosts something. We want to hold them up too...We put eggs away for the annual meetings and the SNEP graduations and preschool at the New Year's [celebration], so we can feed the people that support us all year long." (HP)

With continued strong demand for herring eggs, it is likely that these distribution programs will continue. However, without reasonable opportunity to harvest sufficient subsistence herring egg supplies, there are clear stresses on individual harvesters' and STA's distribution systems.

Institutional Harvest Profile: Hoonah Indian Association and Huna Heritage Foundation Distribution Program

The Hoonah Indian Association (HIA) is the federally-recognized tribal government for the village of Hoonah, Alaska, home to approximately 600 Alaska Natives, primarily Tlingit. The Huna Heritage Foundation (HHF) was established in 1990 by the Huna Totem Corporation as a non-profit heritage organization "to foster and support educational and cultural opportunities for shareholders and descendants." Beginning in 2018, they teamed up to provide support for the volunteer boat captain, Vernon Hill, who for many years has been taking a crew to Hoonah on his boat, the *Shirley N.*, to harvest herring eggs for distribution to the Hoonah community. The tradition of Hoonah fishing boat captains doing this dates back to at least the mid-twentieth century.

Vernon typically harvests between 5,000-20,000 pounds of eggs on hemlock trees, the trees typically being cut in Hoonah and prepared for deployment during the trip to Sitka by his crew of two or three helpers. On some occasions, however, he has not been able to harvest sufficient eggs because of poor spawning or theft of branches. In some such years, members of the Sitka Tribe have helped to supplement Vernon's harvest, so that he returns to Hoonah with a respectable load of eggs for distribution.

An estimated 100-200 residents gather at the dock for the distribution when Vernon's boat, the *Shirley N.*, returns to port. Announcements are sent out on Facebook by the tribe, heralding the time of his arrival. Vernon prefers not to restrict people in terms of the quantities they take, so David See, as community development director for the tribe, steps in to perform this important allocation role to insure equitable distribution. In order to satisfy the most tribal citizens, David must determine the appropriate size and number of bags that participants may fill up with eggs, which are usually laid out on a tarp. Residents may act as proxies for other families, but they are asked to use bags of a certain size (say 8 or 13 gallons if a smaller harvest; or 20 to 30 gallons if it is a larger one), which may be supplied by the tribe. This portion control allows eggs to be distributed to basically everyone who wants them. David See reflects: "And you know it... it's a shame that we have to [limit it] as far as the distribution and the quantity. You know it's a way for me to get it out there so everybody in the community gets eggs. And it's—I say community, it's the community as a whole. It's not just for tribal members or Natives, it's for a community..."There's both Natives and non-Natives down on the dock" (DS).

David See estimates that HIA provides eggs to thirty seniors, approximately five or ten disabled community members, and then up to hundred or more able-bodied individuals who come down to the dock to collect eggs for a total of 140 households: "That is almost every household in Hoonah" (DS), or at least every household that wants eggs. Typically 200-400 pounds of eggs also may be set aside for the Hoonah School *ku.éex'* or "invite" party at the end of the academic year. This is a major cultural event, feeding some 300 people. Some eggs may be reserved for a memorial *ku.éex'* or potlatch, although the latter are often acquired directly through relatives in Sitka. (For example, Wade Martin, a descendant of Hoonah, who lives in Sitka, sent eggs for his father George Martin Jr.'s memorial *ku.éex'* in 2016).

Collectively these types of cultural events might involve distributing eggs to another 300-1,000 people, typically in small portions, depending on how many events are scheduled during a given year. If we consider these portions set aside for communal events, Elders or those disabled, combined with the secondary distributions that occur after the 100+ people have collected their portions from the *Shirley N.* and shared them with family and friends, it is likely that everyone in Hoonah who wants eggs, be they Native or non-Native, is able to get at least a small portion. These portions do not always meet their needs and may fall well short of customary and traditional levels, but still represent an equitable distribution of the total harvest.

Vernon Hill does not request money and is not paid for his services. He emphasizes why volunteering is important and embodies the Tlingit subsistence ethos:

> "[T]he reason I'm doing it is because I really feel that we should go get our own, even if it's a long trip like that. There's always been somebody from here that was willing to go down and get it and bring it back, you know. But then to have it delivered like that and somebody making money off it is a totally different idea than subsistence." (VH)

HIA, in recognition of the value of the harvest to the community and the significant costs involved in delivering eggs, has for at least five years supported the effort by offsetting Vernon's fuel costs. This year, due to a tight budget, they teamed with Huna Heritage Foundation, which covered Vernon's fuel expenses of $884.73 (diesel fuel and some oil), according to Amelia Wilson, executive director of HHF.

The Hoonah Tribe does not have a full-scale tribal foods program, like the Sitka Tribe of Alaska, but herring eggs carry a similar significance. Thus, they support the tradition of Hoonah fishing boat captains who have brought herring eggs from Sitka to Hoonah for many decades. Vernon sees himself as following a tradition laid out by prior seine boat captains, including his older brother Paul Rudolph and his uncle John Hinchman, and others who would go to Sitka each spring to gather herring eggs for the Native community in Hoonah. These two communities are closely tied, including through intermarriage. Related Raven clans (e.g. T'akdeintaan, L'uknax̲.ádi) and Eagle clans (e.g., Kaagwaantaan, Chookaneidí, Wooshkeetaan) inhabit both Hoonah and Sitka, and travel, trade, and interaction between the two communities is longstanding and frequent, especially for fishermen. One of Vernon's crew in 2018, Andrew Larson, is originally from Sitka, and has harvested herring eggs there since childhood, and even crewed in the sac roe seine fishery (which he views as unsustainable).

According to David See, "This whole process of gathering herring eggs starts from the community. … prior to Vernon going down south [to Sitka] he has different community members to help him gather the trees, so he can actually start preparing those on the way down to Sitka, so he's not going down there and looking for trees. So, the process starts [here]…well before he takes off." Typically, Vernon gets information about the timing, conditions, and location of the spawning from local experts in Sitka, especially Mike Miller, while also following the commercial fishing notices. According to See, "typically, it's when they get the two-hour… notice for seine openings that they start getting kind of interested in going down [to Sitka]."

In addition to tribal notices of Vernon's arrival on Facebook, recipients also share their anticipation and thanks. Amelia Wilson quotes a Facebook post: "When it's herring egg season, everyone will be soon gathered at the dock waiting for the arrival of the boat bringing herring eggs in. It is a wonderful thing that is." Another convivial post reads: "My happy place, sitting at my kitchen table with a shot glass of seal oil and soy sauce for dipping my fresh eggs in as I cut branches into smaller chunks (laughs)." There are also messages of gratitude: "Thank you [to] the Indian Association, Captain Vernon Hill and crew for bringing herring eggs to our community of Hoonah" (AN). Individuals collecting eggs may also give Vernon and his crew small donations (cash or gifts) as tokens of appreciation for his efforts, though these are neither expected nor solicited.

The Hoonah community herring egg distribution program is uniquely effective, equitable, and efficient, and generates a host of important subsistence values to the community. These range from food security and first fresh fish of the season, to community spirit and conviviality, to the perpetuation of key skills, traditions, and values among boat captains and crews, tribal leaders, Native organizations, and citizens that participate in their harvest processing and consumption, and, finally, to the celebration

Vernon Hill in the wheelhouse of his troller, the *Shirley N.* He and his crew typically spend up to several weeks in Sitka Sound in March/April monitoring herring returns and collecting herring eggs on branches to bring back to the community of Hoonah. Sitka Sound is the last remaining stronghold of spawning herring in the region and the only significant supply of eggs on branches available to subsistence users in Sitka, Southeast Alaska, and beyond. (Photo: Bethany Sonsini Goodrich, Sustainable Southeast Partnership)

of Hoonah's heritage ties to herring and to Sitka. The benefit flows of the distribution significantly touch nearly every resident in the community. At the same time, the distribution system is vulnerable, with only one boat going to Sitka. The crew has to be at the right place at the right time to capitalize on herring egg harvest opportunities. Other risks to the Hoonah community distribution system, such as rising fuel costs, have been effectively mitigated through collaboration and reciprocity between tribal organizations in order to optimize the benefits of egg distribution.

Institutional Harvest Profile: Southeast Herring Conservation Alliance, Sitka

The Southeast Herring Conservation Alliance (SHCA) has as its purpose "to protect and enhance access to the herring resources of Southeast Alaska for the benefit of Southeast Alaska sac roe herring (G01A) permit holders" (https://www.guidestar.org/profile/26-4581881). They are known as the "industry group" by many Southeast Alaskans and generally advocate against subsistence measures that would restrict commercial fishing activities. They also offer proposals that they see as improving fisheries management.

One area they have identified for improving management is to get a better definition of the "amount necessary for subsistence" (ANS, discussed above), which, if arbitrarily set too high, could impinge on commercial fishing opportunities, should the need for allocation among user groups arise.

Another goal of SHCA is to improve the distribution of herring eggs, as the organization believes there are enough herring eggs out there to feed the demand, if the distribution system can be made more effective (according to SHCA representative Steve Reifenstuhl's 2018 testimony to the Alaska Board of Fisheries). To this end, they collect money from the seiners ($1000-$2500 dollars a boat; the exact amount is not disclosed) to support one or more boats to harvest eggs on branches and return them to the docks for distribution. In effect, SHCA operates as a producer-distributor charity, but somewhat differently from STA.

At the dock, eggs are collected, weighed, recorded, and then distributed to anyone who wants them with reportedly no limitations on how much is taken. The information recorded includes weight, which they use to help build a profile of the "amount necessary for subsistence." Eggs are distributed without discrimination and recipients are not recorded or prioritized. In some cases, in the past, Juneau Natives, perhaps even on behalf of an organization such as ANB or another, have travelled to Sitka by ferry with a truck to load its bed with eggs from this distribution.

The egg harvesters are professionally employed, earning tens of thousands of dollars (exact amount is not disclosed), depending on the tonnage they bring in (paid at perhaps $1-2 per pound). The program has been running for about twelve years and generally gives away between

5,000 and 25,000 pounds, although it varies and could be considerably higher in some years.

Recipients are grateful for the donations, and there is a reluctance to publicly criticize SHCA for charitably giving herring eggs away. However, most harvesters are skeptical of the industry's motives in promulgating the program, as noted in the above discussion. The industry is seen as trying to limit or control subsistence by: 1) measuring supply to establish a lower minimum amount necessary for subsistence (so commercial fisheries will not be unnecessarily restricted in their allocation); 2) demonstrating their charitable giving to the community; and 3) proving through efficient delivery of egg supplies that the sac roe fishery does not present a conflict with subsistence uses.

All of the major harvesters we interviewed see the SHCA distribution program as an attempt to whitewash the real damage that has been done to spawning herring stocks by overfishing for herring roe in areas vital to subsistence. They also feel they are suffering from direct competition from industry-hired egg gatherers, who set early in the best places (as they are employed to do so and want to maximize tonnage), thus crowding out local subsistence users.

Due to crowding out, there is also a potential diminishment of enskilment—harvesters enacting and sharpening the suite of skills that inform the integrated knowledge-practice-belief system underlying herring egg harvesting. Finally, with the loss of skills comes a crowding out or diminution of the values and benefits associated with customary and traditional herring distribution. Harvey Kitka observes,

> "[I]f you just provide it [eggs] to everybody who wants it then you're going to be, you might say, undermining the learning, and the sort of stimulus for people to learn… [to get herring eggs]. [T]hat leads to the question of the industry program...They do not discriminate, right? So, they go get the eggs, they pay people to get the eggs, and then they come back, and they weigh them. But then they're placed out and anyone can come down and get them. There's no priority." (HK)

A minority perspective is that the industry should be providing eggs and distribution services for the subsistence communities "as a way of giving back [to them]. They're the ones taking all the herring, so [they should give back to the subsistence communities]" (AL). This compensatory perspective was not well-developed, however, and only voiced by one respondent.

Most local egg harvesters found the SHCA program of paying people for egg tonnage objectionable:

> "But what's going on today, and I kind of reluctant to say, these guys are harvesting, the boat that's harvesting now, and they're getting paid for it. They're getting paid by the pound. They wouldn't allow us to do that. But for those herring fishermen, they're allowing it. You can't just make a law and then ignore it." (CS)
>
> "And I don't know how exactly how it goes, but there's like a co-op of seiners, ...That [sac roe seiner industry] -- a person here in town hires people to ... get herring egg tonnage for them to distribute to the communities to make themselves look good. At least that's the way I see it, to cover up the fact that they're overharvesting herring. It's kind of disheartening." (CM)

Beyond doubts about the industry's charitable motives, there is a question as to the legality of the harvest itself. In its early years, SHCA hired an out-of-state Native to carry out the harvest, which the tribe protested was illegal (due to the individual's non-Alaskan residency). However, the broader legal question of whether the industry's paying of harvesters to deliver herring eggs for subsistence users is in fact a commercial, rather than subsistence, activity has not been addressed.

Institutional Distribution Profile: Sitka Senior Center

The Southeast Alaska senior center at Swan Lake in Sitka participates in the distribution of herring eggs through its role as a hub and gathering space for seniors 60 years of age and older. Supported by the Catholic Community Services network, the center describes their mission as follows:

> Many older adults and their families are struggling to bridge the gaps between work and retirement, full independence and limited support, good health and chronic conditions. Senior centers are instrumental in bridging those gaps. Local senior centers have regular operating hours and are often the community's focal point for older adult services. The senior center provides excellent opportunities for socialization, relaxation, and participation in a variety of programs and activities. Senior centers throughout Southeast Alaska offer a wide range of programs and activities for older adults, services which may include congregate meals, transportation, education, counseling, activities, and health screenings. Senior centers and the service programs they offer provide a link with local community resources available to older persons (https://seniorcenter.us/sc/catholic_community_service_swan_lake_senior_center_sitka_ak.)

The Swan Lake Senior Center serves lunch to approximately 30 seniors each weekday at its facility, provides meals to homebound seniors, and serves as a gathering space and focal point for health screening, transport services, and education in the community. Lee House, the local director, reported that the center's policy prevents them from serving donated herring eggs for their luncheon meals. However, the center does accept donations of herring eggs, which come to them each spring from both individuals and STA, which they will help process and make available to their seniors. Processing may involve packaging or repackaging herring eggs on branches into bags (or smaller bags), facilitating equitable distribution for seniors to take home with them after their midday meal (or to deliver to homebound seniors with their meals). Thus, the center acts as a conduit for distributing bags of herring eggs. Individual donors often prefer to remain anonymous, Mr. House reported, though some do not mind being acknowledged.

In 2018, a poor year for subsistence herring egg harvests, Mr. House estimated that approximately 30-50 pounds of herring eggs were received and distributed to about 80 percent of the 30 or so seniors that come to lunch at the center, with each Elder receiving a quart bag. In better egg harvesting years, typically one or two gallon bags are distributed. An Elder collecting on behalf of another may request and receive an additional bag. According to Mr. House, who has been involved in these distributions at the center for more than six years, the Elders are "always excited" to see the eggs and "always thankful" for the donations, though some wish there were more, as the spawn used to be more abundant.

Institutional Distribution Profile: Sitka Salvation Army

The Salvation Army is a Christian mission that operates a soup kitchen and other facilities on the corner of Sawmill Creek Road by Swan Lake. According to its website (accessed December 2018):

> "The Salvation Army, an international movement, is an evangelical part of the universal Christian Church. Its message is based on the Bible. Its ministry is motivated by the love of God. Its mission is to preach the gospel of Jesus Christ and to meet human needs in His name without discrimination." (https://www.salvationarmy.org/ihq/Mission)

At its Sitka venue, the Salvation Army provides a free lunch every weekday, Monday through Friday, from 11:30 a.m. to 12:30 p.m.

When we visited on July 23, 2018, David Kitka was cooking pork and preparing coleslaw and other dishes for an expected 45 diners for lunch that day. On a busy day, the number of luncheon guests can exceed

90. Native food meals are generally featured on Fridays. These special meals may include fresh salmon, fresh herring, deer meat, smoked fish (salmon, steelhead), fish heads, and even smoked black bear (up to six times a year). The soup kitchen does not advertise these meals, but they are posted organically through Facebook and perhaps other social media. Food donations and gifts are received, and thanks are given in the form of prayer (grace) that precedes the meal. Donors may be recognized. Ray Nielson was one donor of herring eggs who was mentioned.

The Salvation Army generally does not distribute herring eggs outside of these meals. However, about six or seven years ago, David Kitka recalled that they did package up some herring eggs in bags to send to the Salvation Army Congress in Wrangell. Mr. Kitka also stated that the Sitka branch had sent bags of herring eggs to the Congress in Hoonah this year (2018), although in limited quantities, due to the poor harvest. The Congress typically takes place in early April, thus coinciding with the herring egg harvest season, and is attended by up to 175 people.

Institutional Distribution Profile: SEARHC Hospital, Sitka

The Southeast Alaska Regional Health Consortium, SEARHC, describes itself as "Alaska Native People working in partnership to provide the best healthcare for our communities. Partnering with our communities to promote a healthy balance of mind, body, and spirit." Headquartered in Sitka, and funded primarily through the Indian Health Service, SEARHC is present in 28 Alaska communities in Southeast and beyond, with 100 providers, 1,070 employees, and more than 200,000 patient visits a year (https://searhc.org/). SEARHC's main facility in Sitka is the Mount Edgecumbe Hospital (MEH or the hospital). The hospital is described as: an acute, specialty, primary, and behavioral healthcare facility that provides comprehensive medical services for all community members in Southeast Alaska."

Meals are served to about 100 patients a day through the personal service dining program, "Expressly for You." The cover of their handsomely-produced menu features a photo of lighted boats in Sitka's Thomsen Harbor, opposite the hospital, and quotes the Southeast Tribal Value, "Respect for nature and property; we are stewards of the air, land and sea."

Lexi Smith, the chief manager of Expressly for You, reported that herring eggs are served through SEARHC's dining services and are flagged as a special treat, a "chef's special." This year, as of April 1, 2018, MEH had received one donation of 50 pounds of herring eggs, along with 25 sockeye from Redoubt Bay, mountain goat meat from the Baranof Island hunters, venison, and huckleberries. She noted that the herring eggs received were among the best she had seen this year but were not as thick as in the previous year, her first with the program. When available, herring eggs are served fresh as a weekly treat, two ounces per person, usually as part of a herring egg salad (a means of stretching the eggs into a balanced meal). They might also be served with other Native foods, such as venison stew or reindeer mac and cheese. Alaska Native patients from throughout the region, including Tlingits, Haidas and Tsimshians, particularly appreciate the Sitka Sound herring eggs, because they only come once a year. Donors of herring eggs are usually thanked, either with a write-up in the newsletter and/or with personal thank you cards. Some of the donors, such as Dale Lindstrom, are SEARHC's own employees.

Through their health network, SEARHC also promotes wellness classes, including eating healthy Native foods. A 2017 workshop called, "Herring Eggs and Popweed – Harvesting and Enjoying Local Delicacies," featured Cyndi Reeves of the Wellness Coalition. Such an emphasis on herring suggests that awareness of their health benefits is increasing and could further expand demand for herring products, including subsistence egg harvests from Sitka Sound.

Institutional Distribution Profile: Sitka Pioneer Home

The Sitka Pioneer Home houses 75 Elders, providing care for Alaskan men and women seniors from the level of independence (Level I) to basic assistance (Level II) and onto full-time comprehensive assisted living for Alzheimer's Disease and related disorders (Level III). Meals are served both in a communal dining room, as well as through in-room services.

The Sitka Pioneer Home is also an "Eden Alternative® Registered Home" for end-of-life care, "committed to providing compassionate care to residents, staff, family, and community in a home that nurtures the human body, mind, and spirit while preserving dignity and individuality."

Sitka and other Alaska Pioneer Homes are managed by the Alaska Department of Health and Social Services.

According to Gregory Owens, food service director, the Sitka Pioneer Home typically gets about 8-10 gallons of herring eggs per year donated in boxes or bags. They receive eggs primarily from STA distributions. They serve at least one fresh meal, usually lunch, but the rest are typically vacuum packed and frozen to be served later, usually in a herring egg salad that is served about once per month throughout the remainder of the year. The eggs are blanched as prescribed by the Native way of preparation and may be rendered as salad and put on the salad bar. "People notice," when they are put out, Owens says, and "they love it." Around March, residents will start asking for them, he notes, knowing the herring are soon to return.

Some eggs delivered to the Sitka Pioneer Home are also redistributed to other homes in the region. This year, for example, a gallon bag was sent to the Juneau Pioneer Home. Dietitians on staff (including Melissa Chulpac) help inspect the eggs and plan meals around them. The pioneer homes are catered by NANA (the Northwest Alaska Native regional corporation), which has progressive guidelines on serving Alaska Native foods as a vital and appropriate part of a healthy diet, especially in Native communities. Herring egg donors are thanked for their contributions with letters from the residents. The Sitka Pioneer Home also receives donations of other high-value subsistence foods, such as venison and rockfish. Some of these come through the activities director who helps coordinate with the Taste of Sitka festival event.

When people have access to favored subsistence foods, they are not only healthier but also happier. Owens gets a lot of positive feedback and grateful comments on the provisioning of herring eggs. The residents also have a Resident Council that identifies the cultural foods of highest popularity. Herring eggs are high on this list, perhaps because they are an early spring arrival when "everyone is looking for something to show that winter is losing its grip, and spring has finally arrived. The herring spawn is that signal, a sign of the new life arriving," Owen observes. The arrival of herring also stimulates the Elder residents to reminisce about how things used to be, and they comment on how there used to be more herring in Sitka Sound, with lots more herring in distribution. Residents would welcome more eggs, if they were available, both eggs on branches, which constitute the majority of donations, and spawn on kelp, of which they also receive small quantities.

Institutional Distribution Profile: Sealaska, Juneau

Sealaska is the largest among the thirteen Alaska Native Regional corporations created under the Alaska Native Claims Settlement Act of 1971, boasting more than 20,000 shareholders descended from Southeast Native communities but now residing throughout Alaska and outside the state. In addition to managing or investing in numerous businesses to provide employment and dividends to shareholders, the corporation invests in the non-profit Sealaska Heritage Institute (SHI), student scholarships, and other programs and causes that align with its mission of promoting sustainable socioeconomic benefits for its shareholders according to Southeast Native cultural values.

Sealaska distributes herring eggs, mainly to Elders in Juneau and other communities, who may not have other sources of eggs. The practice of distributing eggs began approximately twenty-five years ago with Clarence Jackson, a long-time member of the Sealaska board of directors. Clarence Jackson used to take his boat over to Sitka to harvest eggs, the majority of which he would distribute to friends and relatives. Sealaska began to sponsor his fuel, food, and accommodation in Sitka, along with helpers if needed, in return for their gathering and processing eggs for distribution. The harvest was typically carried out in consultation with local experts in Sitka in terms of when and where to go. Processing was typically done in Sitka with Clarence and his assistants trimming up eggs on branches and (less often) spawn on kelp.

According to SHI employee Donald Gregory, who was sometimes deployed to assist Clarence Jackson, the number of 50-pound boxes filled with harvested herring eggs and shipped via airplane to alternate locations was typically more than 100: "I don't ever think I remember it being 200, but I always remember it being over 100 boxes. You know sometimes it was 120...maybe 150."

> "Every time they had eggs, if you were an employee or you had friends that were employees, they usually shared. And this went on way back before Clarence was the one

> doing it by himself, you know. I believe there was somebody else before him that they would just send over or cover their cost for the fuel, for a hotel and meals." (DG)
>
> "[T]hey [Sealaska] had some that went to Ketchikan from there. And there was somebody in Ketchikan like the KIC group would distribute it among their office." (DG)
>
> "I was paid my regular hours to be over there [in Sitka]. And so, I just worked a regular work week but Sealaska Heritage donated me as an employee to go help out with the harvest of it. … He [Clarence Jackson] would organize it, he would arrange on the Sitka side, our transportation if we didn't have his boat, we would have like Danny Marino's boat. We had different people, the Martha K [Herman Kitka Sr.'s boat]. He'd always arrange the way that we would be able to go out and harvest it, but we all stayed in the hotel. Got up every morning and just worked and worked and worked and then 'til we harvested it, boxed it, got it to the airport and then I would fly back to Juneau, basically." (DG)

According to Gregory, STA and Sealaska basically served different constituencies in distribution:

> "Sitka Tribe had boats going out and getting it and they were the ones who came back to the dock when the locals wanted to get eggs. They would allow them to come down and take so much. … but if there was somebody that Clarence knew or a family member of one of the workers would come down and say, 'Hey, can I get a little bit?' then they would give them a box or something, you know. So…it wasn't all just for Sealaska … Some of it was for the crew or family members of the crew that had been out doing the subsisting." (DG)

Sealaska eggs also served a wider but more urban-oriented distribution:

> "[W]hen we shipped it out of Sitka it was going to Ketchikan, it was going to Seattle, it was going to Anchorage and Juneau... [I]t never had to be frozen unless you were sending it to like...to Fairbanks or something and you'd say, 'Freeze that one.' But it would only be one or two boxes that ever got frozen; the rest everybody got fresh." (DG)

Once the eggs arrived at their destinations, there would be people ready to pick up the boxes from the airport and redistribute them to their respective communities.

Of the average 100 boxes of eggs that Clarence Jackson's crew would harvest each year, approximately 40 would end up in the basement at Sealaska:

> "In the garage with all the directors' names on it, all the office workers would have [some, and] they might give us four or five boxes for SHI to split up... So, they would just have gallon bags and snippers and you just filled up what you wanted. Everyone took some first, and then if there was any left you know, anybody else could take more. But you know everybody had to have a chance of getting a little bit of it." (DG)

This would translate to approximately 50-60 immediate staff members within the corporation and Heritage [Institute] taking portions of eggs, "And then ... anybody else that lived in Juneau that was ... associated ..." (DG)

Clarence Jackson would have a list of people to contact once herring eggs arrived: "He had a list and he had their phone numbers. And so, I would write out their name, where it was going and the phone number when it arrived. ... I would say there might have been at least 20 ... that I listed up that—you know, like his family members or like my mom, my auntie." (DG)

Maybe three or four boxes would be sent to Sealaska Timber Corporation in Ketchikan:

> "[I]f there was ever a box left and we need to get rid of it … I'd just call a few friends. And even those guys would take it and give it to their uncles, or aunties, or grandparents. And it was just like a chain that just kept going until it ended up on someone's table being eaten. But you know I would say like from distribution for Sealaska or Tlingit and Haida or, you know, ANB, it probably went a couple more levels down before it was being used. So, it wasn't just the amounts that we were getting that was appreciated, it was appreciated down the line." (DG)

Distribution included Sealaska offices in Seattle:

> "I think some went—not a whole lot because those guys at the Seattle office don't usually eat it … like when [former CEO] Chris McNeil was working down there, of course they sent some down to him and he would distribute it amongst … the crew that worked down there in the Seattle office…And I can only remember one time, one time that we had too many and we had to come up with some names to ship to. Only one time out of all the times…" (DG)
>
> "[But] we could never give everybody some who wanted some. … [T]here was always some shareholders [who objected to the distribution]: 'Oh! That was my money…' [i.e., so I should get some, or it should be used for financial investment or dividends, not cultural ones] … So those people I didn't give a second care to. But if a grandma called up and said, 'I haven't had any [eggs], could you help me find some?' Oh yeah, I would try my best and, you know, but for the most part, we could never even come close to serving the shareholder base. If we tried to do that, man we would have to have five seine boats gone constantly!" (DG)

Sealaska would distribute approximately 100 boxes of herring eggs of various sizes each year:

> "[S]o I mean the boxes is the main number but there may have been like 80 boxes of it were the smaller boxes. And then some of them were the great big boxes..." (DG)

Each of those boxes would then be subdivided once it reached the recipient:

> "[S]o if it came to a director, that went to a director's family. And like for us [SHI], we got six boxes and that went to everybody in our office. And then we would, you know, like I'd give some to somebody who didn't have any. For me, I would take whatever I wanted just for eating right now, and the rest I would just let them have at it because I know I have a supply...Those people who are getting it, they did something to help out and sustain the corporation." (DG)

Sealaska's program was mainly internal for its employees and select shareholders or associates, in part because there was not enough to give to all shareholders, but also because they did not want to be in the distribution of eggs business. Thus, Gregory would steer interested parties to other local Juneau organizations that were fulfilling this more general distribution role: "If someone called up and said, 'Hey, you guys, Sealaska's giving away eggs.' I'm going to tell them, 'No, they're giving it away at ANB today. Or at Tlingit and Haida you can get a grocery bag full. They're giving everybody one grocery bag.' Just distributing the information." (DG)

CHAPTER 6: BENEFITS OF DISTRIBUTION

"What really boggles my mind is that the majority of the resource is being overfished for another country. It's not even for our own use."

—Tammy Young, Sitka

Photo: Bethany Sonsini Goodrich, Sustainable Southeast Partnership.

Food Security and First Food

> [Herring spawning is] kinda like the beginning, the rebirth of our ocean every year. It was just amazing out there, I love it out there that time of year…
> —Wade Martin

Although historically herring were not the largest segment of Sitka's wild food production—that distinction belongs to salmon—they were integral to food security because of their early spring appearance and their foundational role in replenishing the Sitka Sound marine ecosystem. Herring were the first major fish population to return to Sitka Sound and other areas for spawning in the new year, and they drew other important subsistence species back with them, including salmon, halibut, and other fish, as well as sea mammals, such as harbor seals and sea lions (and humpback whales), and masses of seabirds. Thus, abundant herring returns are highly correlated with food security and the renewal of the annual cycle of food production.

As Heather Powell attests, food security is ensuring your family has what it needs:

> "When you go out and you drop trees, you're doing them with the intent that you're gathering you know, only what you need. But as a family, your needs are larger. Your needs are to be able to give to our grandparents that no

> longer have their husbands, you know the widows in our community. To make sure that they have what they need to be strong. And by strong, you're strengthening their soul with those foods." (HP)

In the era before freezers, the later winter period could be lean, so herring eggs were not only eaten fresh when they were caught in the spring, but also preserved by drying:

> "Well, a lot of people didn't have freezers… [and] there wasn't that many jobs in the community except for fishing back then. So that was just one of the things that we saved for—we put up for winter. … I don't think each family got more than 50 pounds [of herring eggs]. … Some of the seine boats went down to Sitka [at least two of boats]." (GJ)
>
> "[Y]ou'd air-dry it. They used to hang it in the trees and air-dried. So, for generations after generations that's been a staple of being able to eat herring eggs. And we don't just eat the herring eggs, we eat the herring too, you know, if we can get the fresh herring." (DG)
>
> "And back then we didn't have any freezers, so my mom would dry it. And she would always say, this is what we're going to use for the winter. But I came from a large family and my mom couldn't dry it because we were always eating it as fast as it would dry [laughs]…She didn't dry it outside. We had an oil stove in the kitchen and she had some line and she wouldn't make the branches too heavy. Just enough to hold [them] up there and just let it stay there for a couple days until it was dry. And then they would pack it away in boxes so that you know the bugs or whatever wouldn't get at it. But that's how we saved, you know, so that it didn't go to waste." (DG)

Drying eggs also made them more portable, due to the loss of weight with dehydration, and the fact that they could be easily packed for transport. This allowed for (re)distribution of food in time and space.

Melody Leask reported that "[Herring eggs] would be the first [communal subsistence] meal. It would be something that we would celebrate because we made it through the winter, and our harvest season is here" (ML), while another harvester, Mike Baines, commented:

> "Well I mean one of the things I think about is ... it's like the beginning of a new year for me because after a long winter went on and deer season's been over for a couple months ... it's kind of an exciting way to start the new year of gathering stuff. You know, and then after a while would be seaweed and, oh look, there are gumboots, and that kind of thing. So it's kind of a good way to start off the new year for me anyway. It's kind of a boost to peoples' morale I think." (MB)

Amelia Wilson of Hoonah corroborates this sentiment:

> "It was just… fun and exciting. It was spring, and … you knew that the foods were available at this time. And of course, everything is better fresh, so I think that was just the excitement of something you don't get regularly. I think like fruits used to be maybe, nowadays you can get fruit almost any time of year, but it used to be you'd only see the mandarin oranges during winter time. ... So, there was just that association of you know, 'It's spring! And there's herring eggs! Yay!'" (AN)

Tom Young adds,

> "I just couldn't imagine spring without it. I mean, that's—it kicks everything off. It's how we know that, you know, the herring eggs are there, and we start looking at how healthy the gumboots might be. The next thing you know, we're on to seaweed and…beach asparagus, goose tongue, I mean…it's just—it's the beginning of a very enjoyable path through the season." (TY)

The boost to peoples' spirits, in turn, sparks communal gatherings of convivial feasting, which is discussed in more detail below.

Another way that herring contributes to the benefits of food security is by being reliable. While herring spawns fluctuate, prior to the reduction era, local and traditional knowledge suggests that they were always superabundant and thus reliable for harvesting. Nowadays, the security of supply has become less consistent because spawning has become spottier and shorter in duration and extent (Thornton, et al 2010a).

Moreover, the pressure that the sac roe fishery puts on spawning herring can affect reproductive security. Albin Frederickson notes:

> "[T]hey [sac roe seiners] mop the herring up where they can spawn because that's what they wanted the herring for is for the roe, you know. ...if they didn't have such a big fishery and give them another 2-3 thousand tons of herring, everybody would have done a little better and had a little bit more for home and for the freezer and whatnot. But as far as the fishery goes, like I said, I'm a commercial fisherman. I fish, I understand, you know, a person's got to make a living. And, but the one thing about subsistence herring eggs—we're just taking a few eggs. We're not killing the herring. They can come back and spawn for another ten years, you know." (AF)

When asked what it would mean to him if there weren't enough herring to harvest eggs and distribute them, Tom Gamble, laying blame with the sac roe seiners, replied: "I'd probably be an owner of at least 52 new seine boats. ... I would sue them all. I would own every fishing boat that ever ran. ... That's how important it is. ... They [the sac roe industry] took basically an identity, saying I couldn't be a [Kiks.ádi] person anymore." (TG)

> "[Herring] has been around for generation upon generation and I can't see Sitka and [with] no eggs in the future ... It might be different for Craig and West Behm Canal, all those others to shut down, but for Sitka to

> shut down is significant. It's a significant event [for all] Lingít Aaní." (TG)

In general, food security associated with herring has declined. Spawning areas, spawning abundance, and deposition quality (thickness) have overall decreased and become more variable. Some question whether it's worth the effort:

> "You got all that physical labor of getting the trees. You got to get your fuel and maybe if you have engine costs or whatever, and then taking the time off to go do it. Then you have to deal with the weather. Sometimes the weather is an issue because it gets a little bumpy out, nasty out. And then you do all that work and you're chasing them around and you put a lot of money into it and then you're anticipating a good return [of spawning herring] and it just never happens. And then you're like, 'What the heck?'" (CM)

Health and Wellbeing

Herring eggs, meat, and oil are well-known for their health benefits, as are sardines, which are sometimes cross-packaged with herring. According to Dr. Andrew Weil,

> Both herring and sardines are very good for us because they provide high levels of heart-healthy omega-3 fatty acids, essential fats that also play an important role in normal brain development and function. Research indicates omega-3s down-regulate inflammation and may help reduce the risk and symptoms of a variety of disorders influenced by inflammation, including heart attack, stroke, cancer, and autoimmune diseases such as rheumatoid arthritis. People with high cholesterol, diabetes, symptoms of premenstrual syndrome, coronary artery disease, breast cancer, memory loss, depression, attention deficit hyperactivity disorder, insulin resistance, and

> arthritis may also benefit from the omega-3 fatty acids found in fish oil.
>
> Another plus: because they're relatively small and near the bottom of the food chain, herring and sardines don't accumulate the contaminants that are so common in large, predatory fish…
>
> Fresh herring is higher in omega-3s than herring sold any other way... (Weil 2018: https://www.drweil.com/diet-nutrition/nutrition/herring-or-sardines/).

Herring eggs are higher in protein and lower in fat than herring meat or oil but also contain fat and important nutrients, such as vitamin A and C, calcium, and iron (see Figure 2), as well as selenium, a powerful antioxidant.

Sitka egg harvesters attest to these nutritional values and other health benefits, especially for Elders. As Mike Smith states, "Herring are a sign of … health and renewal. It kind of sets off everything after that," adding:

> "I think it really supports the Elders… it helps out people that can't go out and harvest. And it really helps out people that are low income and we see them quite often. And it kind of makes you feel good when you can help them out by giving them traditional foods. And you know, it's not only healthy and high in proteins and all that, amino acids and stuff, it's a good feeling to be able to help them." (MS)

Jeff Jackson's wife works for the Alaska Native Tribal Health Consortium (ANTHC), which advocates for Native foods, such as herring eggs, to be available in hospitals and other health institutions:

> "She used to run a program called 'Store Outside Your Door,' which promoted healthy living with you know, foods like herring and other things, herring eggs. And so, they did a piece on that and a bunch of people … came down [from Anchorage] and harvested with us… I got a lot of 'thank yous' and I got a card. … From one of the people that came down on behalf of who they work for…

[T]hey paid for everything, the boxes and everything, including the shipping of eggs back to Anchorage. I think they incorporated it into their meals at the hospital." (JJ)

Jackson also sends eggs to Kake, Anchorage, and other places, like his father, Clarence Jackson, did on behalf of Sealaska—collecting eggs to distribute to Kake and other communities:

"I just, I stepped it up from what we did with my dad and actually had a crew. And we put more trees in the water. And we were bringing, we brought loads of herring eggs just to the community of Sitka and distributed there. And sometimes I would, well I boxed up some for whoever wanted some. And I think...I think it was 2016, or '17, I'm not sure, we sent a bunch up to the hospital in Anchorage, the Native hospital. ... [B]etween 10 and 15 boxes." (JJ)

Beyond individual health is the notion of communal wellbeing. If herring benefit local species' health, including human communities, why export it to other countries in way that might jeopardize the wellbeing of the local social-ecological system? In this respect, the status of health and wellbeing becomes a resource management question.

"What really boggles my mind is that the majority of the resource is being overfished for another country. It's not even for our own use. ... if people could stop and think about that for a moment... What is the purpose, when it's not even to help our own children or our own Elders be healthier or more productive? I mean, not that we shouldn't be contributing to other peoples' wellbeing, but it just seems as though, if you use your logic, it doesn't make sense that we would overuse a resource for another country that doesn't really know that this resource could go away. Otherwise they wouldn't continually be utilizing it the way they are." (TY)

Wade Martin contrasts the feeling you normally get from a bountiful herring egg season to the depressed situation now, which corresponds to the antithesis of wellbeing.

> "The feeling you [normally] get when herring season comes around, you're just like a kid in Christmas and you can't wait to be a part of it and watch the eagles, everything around and the whales and all the marine mammals and now… and now you go out there and… it's desolate. It's lifeless. The powers to be are lying to everyone about the health of the stock, they're lying about the [amount of] spawn, you know, it's just a big lie anymore and it's disheartening." (WM)

He also takes issue with the way the Alaska Board of Fisheries makes its decision about herring as being counter to the wellbeing of the community:

> "Board of Fish meetings and get all these monkeys to speak what they want them [to], you know, a script. Saying 'Oh yeah, everything's healthy and you're [subsistence users or conservationists are] gonna ruin our life because we're not gonna be able to kill these fish and make a living. You're impacting my living…' [I]t's very well-orchestrated, very well organized; it's big money. A person like me doesn't stand a chance going into a board room like that, a stacked deck, and expect to come out with a result you're happy with. You're basically giving your testimony and they say, "OK, thanks a lot, now get the hell out of here, and we're gonna do what we're gonna do." That's the way it's been for a long time. I don't see that ever changing." (WM)

This sense of powerlessness over unhealthy management becomes dispiriting and undermines community wellbeing.

Nutrition Facts

Serving Size	100 g
Amount Per Serving	
Calories 74	
% Daily Value	
Total Fat 1.9g	**3 %**
Saturated Fat 0.4g	**2 %**
Cholesterol 40mg	**13 %**
Sodium 61mg	**3 %**
Total Carbohydrate 4.5g	**2 %**
Protein 9.6g	**19 %**
Vitamin A 1 % • Vitamin C 1 %	
Calcium 2 % • Iron 15 %	

Figure 2. Herring egg nutrition facts, based on 100-gram portion (2,000/day calorie diet)

> "[We] sit there and [they] give us three minutes to spill our guts then dismiss everything we tell them...it's just a process that they have to go through to achieve their goal and maintaining what they want and what they want to keep. So, I guess in our own way, we're indirectly a pawn in their game and they need us for that but anything else and they don't want to hear from us. They don't care what we have to say. We're the enemy. We want to do away with part of their living and sometimes uh, sometimes things are more important than money." (WM)

Abel Ryan of Metlakatla had this to say about the psychological effects of reduced herring eggs on community wellbeing and cohesiveness:

> "In the short term, the reduction is still having the same [negative] effect because everyone's mind is still on the herring eggs. And while we're not celebrating as much as we used to, the joy of receiving the eggs, there is definitely a larger trend of everybody being angry together about not getting them ... And that is also one of the few things that everybody can get together about and be angry about now too, which isn't that healthy—[not] as healthy as receiving the eggs, but they're still coming together on that recognition. ... It's kind of a group mourning." (AR)

Tammy Young sums up the foundational role that herring play and the benefits they confer on ecosystem health and wellbeing.

> "What I remember being young and listening to them talking was herring...provide the food source for almost every other animal in the water that we gather. They are on the ground floor, so to speak. They are what upholds every other species. And you can choose all the way down to plankton and they are being nourished by herring whether it's the eggs, the milt, or the actual fish." (TY)

One way for the state to manage the ecosystem health benefits of herring would be to recognize it as a forage fish and follow the latest guidelines for management. As one member of STA points out:

> State of Alaska refuses to, and the Board of Fish refuses to, designate herring in Alaska as a forage fish because it would require them to manage differently. And as far as I know…I don't know of another place where there's herring fishing going on that has not, does not consider herring a forage fish because it's the subject of feeding everything in the ocean." (PE)

Managing forage fish generally requires a more conservative set of guidelines than that of the commercial harvest as the focus becomes the conservation of forage fish for their value in the food chain—as the "feed for everything"—and even to restore them to levels associated with healthier communities.

Food Diversity and Status

Herring supports food diversity in two ways: First, by providing an abundance of feed for the marine food web, supporting a rich diversity of critical species at higher trophic levels of the ecosystem, from salmon to halibut to sea mammals to sea birds. Second, by becoming an item of exchange, through barter, trade, and other forms of giving and reciprocity, herring eggs in particular can be leveraged to bring non-local subsistence foods of value (e.g., caribou, eulachon, moose, bowhead whale), as well as other goods (red cedar, inland furs, etc.) to the community. This diversity, based on a herring foundation, has helped make Sitka the largest and wealthiest of the Island Tlingit communities. Herring conferred not only diversity, but also an abundance that could be converted to enhance status and to ensure security and wellbeing.

As Chuck Miller makes clear, there was even diversity and status among herring egg "treats":

> "Oh yeah, my whole life we ate herring eggs. It was a delicacy—well, not a delicacy. It was a real treat when we got it because back in those days it was always super thick and when grandma and grandpa brought them, oh boy, it was awesome. So, yeah, been eating them my whole life except for the last couple years." (CM)
>
> "I would just keep bringing it in until my mom says, 'I think I got enough.' Or my aunt, 'I got enough.' Or, 'Can you get me more?' 'Yes, I'll be right back.' Just go out there, it's like the apples, just pull them up, get them, come back in. Got lots out there, just tell me what you want. And then they would say they want the *né* [eggs on hair kelp], and the stuff on the *daaw* (eggs on macrocystis kelp); they want the *daaw* and the *né*, which is the one on the hair kelp and the one on the broad kelp. But that's usually after the main spawn's happened. Usually, the next big low tide after that is when you go get those." (CM)

Thick herring eggs are preferred, while thin ones are often left to hatch out, thus replenishing the system. And spawn on macrocystis kelp (*daaw*) and hair kelp (*né*) were considered separate foods, with their own taste and qualities.

Seals, sea lions, halibut, ducks, salmon, and other subsistence prey also gathered during spawning time to feed on the incoming herring, thus providing a cornucopia of favored foods. In Kake, it was even reported that halibut, normally a bottom dwelling fish, could be speared when they chased the herring into the shallows during their spring migrations to spawn.

As Hoonah Elder Ken Grant summarizes: "It's an important part for the young people to learn that there's a food chain, and it starts with the little critters [such as herring]." (KG)

George Bennett, an Elder originally from Hoonah but living in Sitka, agrees, pointing out the importance of leaving enough herring in the water, which is recognized in Tlingit law as a means of conserving the diversity of food in the marine ecosystem—in short, valuing herring for its ecosystem services:

> We have our own traditional laws that dictates what we do... And especially the springtime when it comes time to harvesting herring eggs, you know, herring eggs is a very highly prized subsistence food that we would take. But at the same time, we were also concerned about all the other animals and the fish that depended on that to feed them as well. And so, when you look at them diminish and the balance of the ecosystem going astray, you know then that has to be addressed you know. You can't let the resources diminish to a point where you start creating farmed stuff and hatcheries. You know, they're not the same." (GB)

Heritage and Cultural Identity

In 2018, Vernon Hill from Hoonah and his crew cut 200-some Western hemlock trees to haul to Sitka to lay as substrate sets for spawning herring.

> "I've always felt that ... it's pretty important to gather your own subsistence foods. That's part of your identity in a village and being a village and being part of it. And Hoonah's given me a lot over the years as far as teaching me how to fish, old timers telling me things. I figure this is just a way to give something back, you know." (VH)

Vernon's crew is composed of volunteers who also describe the positive identity they begin to embody as a result of participating in the herring egg harvest for the purpose of providing subsistence foods. When Andrew Larson heard that Vernon was in need of volunteers to help harvest and distribute herring eggs, he opted to help out despite no offer of monetary compensation: "Didn't matter. ... I didn't get nothing" (AL). When asked why he would volunteer his time and efforts, Andrew responded: "Herring eggs. ... And this just—I don't know, it's like a trigger to me to know—to go out and get the subsistence food" (AL).

For Amelia Wilson of Huna Heritage Foundation, supporting the herring egg harvest is very much part of the mission of supporting the culture, identity, and heritage of Huna People:

> "[I]t's the first year that Huna Heritage has supported the herring egg collection efforts. And it has been primarily something that Hoonah Indian Association has done. HIA and Huna Heritage entered into an MOU together and that's something that historically, we have worked together as organizations. And so that we just, you know kind of re-committed to the relationship and have been actively meeting on a regular basis to make sure that we're aware of what each other's organizations are doing and how we can be of support because we serve largely the same population, and a lot of our missions and goals and program objectives overlap and so it just makes sense, rather than trying to do the same thing or you know, looking and seeing what areas are served." (AN)
>
> "And so, with the herring eggs, David See, who has been in charge of running that support for that trip … asked if Huna Heritage would be in a position to assist, just because of less funds available all around. …We're governed by a board of trustees. And so, I reached out to the board of trustees, who unanimously supported [it], and of course are all Hoonah connected…. they are all familiar with the annual trip and because there's the inability to collect here. ... And so, to support the efforts of Vernon and his crew to go down means that it's a benefit to the entire community." (AN)

For a heritage organization, like Huna Heritage, supporting herring egg distribution makes sense:

> "[O]ur mission is focused around culture and education and our core areas of service are community cultural programming. Our board has titled it 'Our Way of Life.'

> [Haa Kusteeyí] And so, this is something that they felt really fit into our way of life and providing opportunities for shareholders, descendants, community to be practicing our way of life. And that food is so important and such a big part of that, so that was something that the board felt strongly about supporting and they were all very much in favor of that." (AN)
>
> "[T]hose are the types of things that we'd like to support is—you know our foods are so much a way of our life and so important for a number of reasons: nutritionally, culturally, spiritually. That you know providing access to—ideally if you could go collect it, and if not, you know ensuring that there's access." (AN)
>
> "I think that we would want to continue supporting the trip because then it's not falling on one organization and it's such a wide benefit. It's open to the community. It's very inclusive. It's a means that—I'm sure there are people that get to trade, but I think most people I know, it's their only source, their only option. And so, to me it's something that has such significant value that I would support it and I think that our board would continue to support it as well." (AN)

Similarly, Tom Gamble of Sitka remarks,

> "[I]t's beyond a resource that you just eat and consume anymore. It's... your [life]. What is it to anybody else? I don't even know a single other group of people that can claim a resource so closely." (TG)

One way that Tom Gamble's Kiks.ádi clan identifies with Sitka Sound herring so closely is through *at.óow*, or sacred possessions like Herring Rock (*Yaaw Teiyí*). The Kiks.ádi covenant with herring, as a spiritual bond, is also reflected in sacred possessions, such as the Herring Rock Robe (*Yaaw Teiyí Naaxein*), which functions as a kind of deed of trust to the Sitka Sound herring grounds around Herring Rock. Similarly, young Kiks.ádi women, also known as *Kaxátjaa Sháa* ("Flipping Girls"), are identified with spawning herring at Herring Rock.

Ceremony and Cultural Events

Herring are used not only for immediate consumption as first fruits, and put up for winter supply and trade goods, they are also central to cultural events, including funerals, 40-day parties, *k̲u.éex'* (memorial parties or potlatches), and other social gatherings. In all such events, herring may be served on branches or kelp, or "stretched" to go further in herring salads and other recipes combining them with other foods.

Heather Powell, a Tlingit cultural leader and teacher in Hoonah, observes:

> I grew up [in Sitka] very closely connected to the herring. You know my great-grandfather was Kiks.ádi and the women of the Kiks.ádi [*Kaxátjaa Sháa*], you know they're like that flickering herring on the top of the water. And I have a very strong connection, you know, to that." (HP)
>
> "And so when I'm teaching and when I'm trying to share the little bit that I know, and learning along with everything that I'm teaching, you know to be able to express to them, the stories of, you know, when the Kiks.ádi woman stuck her hair in the water and how that became, how did they know how to do that? You know, the strength behind those things and then to actually be able to see the *né*, and you see the eggs on the hair kelp, and you see the eggs on the bull kelp. You know, and you see the eggs on the hemlocks, and to be able to explain the differences and the strength behind why they use hemlock, and why you'll never see it on spruce. And to let them taste that and feel that, you know. It's a ceremony." (HP)

Ceremonies help to reinforce these connections through connections to sacred foods and to each other:

> "Our world contains so many things that as they start to become less and less, so does our connection with those things you know. And it's really, for me, a really emotional thing. And because it's such an emotional thing, that's when you bring those strengths out. That's when you sing those songs, and that's when you bring your drum and remember that that food isn't only just something that's going to give you sustenance, it's feeding your soul." (HP)
>
> "So, when we talk about what are the benefits of the distributions of herring eggs, this year these herring eggs are going to feed not only our community, but also our honored guests that are coming in that are our Elders that have sacrificed a lot of things in their life in order to continue teaching." (HP)

In Metlatkatla, Melody Leask describes of the communal spirit of having "all the Eagles working together and cooking up big batches of herring eggs (ML)." In Sitka, George Bennett reflects on the difficulties of supporting a cultural party today with herring eggs as compared to the way it used to be even twenty or thirty years ago.

> "And so, we have maybe about, five quarts, bags in our freezer now. I remember, I had one of those long freezers, about this long [5 ft.], nice freezer. ... And I actually filled that thing up with one-gallon bags of herring eggs. And that winter, I went to a party in Angoon. I took 20 bags and fed everybody in that building. And I sent some to Hoonah you know. But I still had those bags in my freezer. But that's how good it was then. Now, I got five quart-bags in my freezer. That's it you know. And the sad thing about it is that, you're going to wait for a special period to bring out that one quart and eat it. This is how bad it's got; it's not where you can reach in there in your freezer, grab a gallon bag out and have a feast every other week maybe, you know." (GB)

Wade Martin displays a good example of thick eggs. (Photo: Bethany Sonsini Goodrich, Sustainable Southeast Partnership)

A major cultural event supported by the Hoonah distribution of herring eggs is a community *ku.éex'* or "Heritage Celebration Potlatch," sponsored by the school. Several hundred pounds of eggs are set aside from Vernon Hill and his crew's harvest for this ceremonial gathering. Daphne Wright, Heather Powell, and other instructors integrate information about the science, heritage, and culture of the herring egg harvest into their curricula. Daphne Wright explains:

> "Yeah, we have herring eggs, right. I think it might have started because we started doing a heritage, we called it our Heritage Celebration Potlatch. We started doing a potlatch. And that was actually 28 years ago. And so, it was—we decided to have a school-wide celebration. We had seen that in Juneau. They had a curriculum where the third grade, I think it was a third grade level curriculum, they did a pretend potlach and they pretended that half the class was Eagles, the other half were the Ravens and they put on a mock potlach. And so, we were looking at that and I did one, one of my years when I was still in the classroom. And we thought, 'You know what? We are Eagles and Ravens here. We don't have to pretend. Let's for real. You really are Eagles. You really are Ravens. Let's teach the kids who they are. Let's have a school-wide celebration.' And so that started. So, this year we're having it again… this is our 28th year; 28th annual … school potlatch. Now we call it *ku.éex'*. We're more correct I think in our Tlingit usage." (DW)

Herring eggs have been offered at the *ku.éex'* for at least the last ten years. The program offers students, who may not be exposed to Tlingit traditional foods at home, the opportunity to connect with herring and other Native foods and traditions:

> "I thought, you know, they need to be introduced to this food that they might not have at home, they need to have it at school too. So that became part of my lessons. And we

would make sure the kids had an opportunity to eat some herring eggs, or even try them, or even see them." (DW)

"[W]e have used it when we have the resource and the opportunity. And it hasn't happened every year, but I know I have pictures of kids that are now in high school when they were in kindergarten eating the herring eggs in their classroom, little paper plate, little bit of soy sauce." (DW)

Daphne Wright tells this story about one of her students:

"Nick Hughes, and he is the great-grandson of Lily White. He was in kindergarten; I was handing out the herring eggs to everybody that wanted some. And at that time, it seems like just about everyone would take some. And they maybe would not eat it, but we would encourage them to take some. And we gave them crackers too in case they wanted to just eat something. There was herring eggs there and a little cracker, a couple crackers. But Nick Hughes, he's looking at this and he's sitting there. I say, 'What's up?' And he goes, 'Where's the grease?' And it just cracked me up. And I said, 'Sorry, we don't have any grease here. You better get some at home. But that is so, he was such a Tlingit boy! That just really impressed me." (DW)

"I show them photographs of a boat—I have a beautiful photograph showing a boat and alongside it is like a tree or a huge tree with, just loaded. ... I also have pictures of the people when the boats have come to Hoonah, when everyone's down there [on the docks]. And there's somebody's uncle holding up a big piece. So, we go through that and talk about it. And I also try to teach them ... the Tlingit word for herring eggs [on hemlock branches] is *haaw*." (DW)

"[T]his year, when I took some in to the K-1 class there were quite a few of the students who didn't want to

> try it, which made me feel bad, but I didn't want to force anything on them. I don't want to have a negative feeling about it, but it made me start thinking, so—one particular child, I know that her father is a very strong, he comes from a very strong Tlingit family, and I know that they do a lot of subsistence gathering, and I know they probably eat a lot of fish and stuff. But the mom is non-Native. So maybe she's the one that cooks? And so maybe this particular family isn't getting [Native foods], you know, [it] kind of depends on who's cooking. But I think a lot of Native men do the cooking." (DW)
>
> "I make sure we talk about it in all my classes [how herring eggs are gathered and transported], so they know what's going on and they appreciate the fact that someone did go over there and gather and brought that here in Hoonah." (DW)
>
> "[H]erring eggs are the one that I spend the most time on. As we're learning food, we're learning for our Tlingit vocabulary". ... And so, as I'm teaching vocabulary, we did a bunch of food and we learn how to say 'bacon,' and we learn how to say, just 'eggs' as in bird eggs, because there's bird eggs and there's salmon eggs and there's herring eggs. And I tell them that you know, you have different kinds of words for these different kinds of eggs. It's not just 'eggs.'" (DW)

Learning in the context of serving the food makes the teaching experience more powerful: "[F]or herring eggs, I got photographs. Yeah, I don't have the real... and that's where tasting it, as we're eating it, yeah" (DW). Daphne Wright estimates that the *k͟u.éex'* benefits the 120 kids who attend the school and their invited guests: "Yeah, so another, maybe another 100, 120, or 130 guests; maybe more." (DW)

While Daphne Wright teaches at the primary school, Heather Powell extends the curriculum to the high school level:

> "I know this year, the high school kids—I think Heather got some from her sister before the big boat came. So,

> the high school kids were over there preparing it, and so we're hoping they get it in time so we can take them to all our classes. And we just didn't get it in time. ... But, our young cultural para-professional worker, I told her, 'Just go in the lunch room and go around with your tongs or whatever, your spoon, and just offer it.' And she said a lot of kids wanted some. So, she just kind of went around during lunch and just went and offered it because I didn't have it for my classes. Yeah, so it's kind of like a snack. And she said quite a few of the kids did [take it] and some kids like it..." (DW)

Heather Powell elaborates:

> "So we have 120 students, we have 30-40 staff, and we have a community of 893 or so, give or take. And you can count on 300-400 people showing up [to the *k̲u.éex'*]. I mean, it's the event. ... [We invite guests]. Last year we had a whole catamaran from Juneau come. ... The entire TCLL, Tlingit Culture, Language, and Literacy program ... and Jessica Chester and Michelle Martin, they brought all their students and all their families...We had an entire catamaran, 150-200 people that showed up. This year we've made special efforts to secure plane rides for Ruth Demmert [from Kake] and Virginia Oliver from Wrangell. She's just recently started teaching high school Tlingit in Wrangell. So, we want to encourage her, we want to lift her up. ... Ruth Demmert just recently retired last year, after teaching over 25 years in Kake." (HP)

Herring eggs are a featured part of this celebration and the context for their presence, including their origins in Sitka, cultural significance, the harvesting efforts of Vernon Hill's crew, and those who processed and prepared them for consumption at the feast, are all recognized, along with the honored guests.

Conviviality, Communion, and Social Capital

The arrival of herring eggs also confers the benefits of conviviality, communion, and social capital. Conviviality can be defined as the positive sociocultural values, wellbeing, and friendly relations associated with local production, distribution, and communal use of local environments and resources. Ivan Illich, in his well-known *Tools for Conviviality* (1973), writes that conviviality involves "autonomous and creative intercourse among persons, and the intercourse of persons with their environment" (1973:24), this being in "contrast with the conditioned response of persons to the demands made upon them by others, and by a man-made environment." Presciently, Illich warned of a "war on subsistence" by the industrial elite and modernization programs which rob small-scale subsistence societies of their knowledge, skills, and ways of knowing place, thus leading not to human prosperity and social happiness but instead to "modernized poverty," dependency, and human subordination to industrialization, efficiency, and growth. Illich proposed a framework that would "invert the present deep structure of tools" for efficiency and productivity towards those that would optimize community and conviviality benefits. For Sitka Tlingits, an abundance of herring eggs optimizes conviviality, while also creating community through the heightened sociality and alimentary communion of eating herring eggs together. This conviviality and communion, in turn, helps to build social capital—networks of trust, reciprocity, and capacity that help society function effectively and resiliently even in the face of social and environmental stresses or change.

Our interviewees spoke of the power of sharing, of giving, and of enjoying herring eggs in communion. As Albin Frederickson sums up, "people are happy to receive herring eggs... Happy, they can't wait to get home and eat some, you know" (AF) with their family and friends.

Conviviality is not only sharing but sharing the alimentary communion (Durkheim 1965 [1912]:378), enculturation, and enskillment that comes with feasting together. As Kiks.ádi house leader Tom Gamble put it:

> "When you really put it into the context of the spring harvest, sharing wasn't like, 'Here, take this bag, go put it in your freezer.' Sharing was like, 'Hey, we're going

> over to Auntie's house [to consume some eggs]. We'll lecture Tommy about what he did wrong on the boat'" [laughs]. (TG)

The last point, though made somewhat facetiously, speaks to a larger point about sharing with Elders not only food but how one got the food, so that the skills of subsistence—e.g., boat handling, setting branches, processing eggs, etc.—can be reviewed for assessment and learning.

Tom Gamble also remarks upon how important it was to Elders and how they always had a special place at the herring egg table:

> "My grandma Helen Kasteen Peters, she was already, I think, approaching 90-something when she was coming to visit us. And as soon as the first boatload [of herring eggs] would come in, she could barely move around. Her hands were really crooked, and she could hardly see, but we told her there were herring eggs out there and she would grab with her hands or whatever she could and just like steady march until she got to the table and then... she'd stay all day." (TG)

In the spirit of alimentary communion and conviviality, the table always included other people. Herring eggs were almost never consumed alone, but around a table, often with newspaper thrown down to make the clean-up of needles and eggs and oil (seal or eulachon) and sauce (typically soy) easer to clean up. Gilbert "Buddy" Hudson of Metlakatla remembers putting down newspapers and inviting "friends and relatives, you know, whoever was hungry (laughs)" over to share in the first herring eggs of the season: "Yeah it's easier to clean up, just throw down newspaper (chuckles)" (GH). Mike Miller reminisced, "Well, I remember growing up...I always knew we were going to have traditional food when I'd go home and see newspapers all over the place" (MM). Ken Grant recalled: "It's that huge pile of herring eggs in the center and then everybody would be grabbing... I don't even think we had dishes. We put newspaper out. It's all over the floor" (KG).

Heather Powell of Hoonah made a similar comment, reflecting poignantly on the sensuous experience and conviviality of her first herring egg feast:

> "You know that food is something you remember the first time you put down a Tlingit table cloth and laid that newspaper across the table and you just sat down with your grandparents and they brought out their best seal grease or their best eulachon grease. And you sat there and ate until your belly hurt. And it wasn't a gluttony thing, it was just the sense of family, you know." (HP)

Jeff Jackson of Kake, who used to help his father Clarence Jackson collect herring eggs and now runs his own boat, reported that the celebrations and joy began even with distributions among the crew, before the boat got to the dock.

> "Me and my dad used to just, we would enjoy it with the people we harvested it with. And I'm guessing that like if we send some [eggs] to Kake [his home village], whoever got it would obviously share it with their family as soon as it arrived.... I've done it for every year and I really enjoy [it] –we pull the first branch out of the water, they'll put it right in the pot and then me and the crew or whoever, we enjoy that first, and then we get back to work." (JJ)

This spirit of conviviality is reflected among other harvesters too:

> "Everybody likes the first taste. Which it's kind of funny to see on the boats too, because it's something I've noticed, and I don't know if people catch this or not, but usually if you're out there's a whole mix of the same people you see that get along. There are some of the newer ones that we're not quite sure, but with the people that reasonably get along is everybody go out and kind of pulling their sets. And oftentimes there will be kind

of little communal gatherings. Little boats pull up next to each other and stuff. And one of the funny things to watch which I've seen is if somebody has eggs, they'll always break off a little piece to hand it over to the next boat to try their eggs. And it's funny because the next boat oftentimes will have a big pile of eggs too. ... And, it's like, 'Oh try my eggs.' You know, and a lot of people do that ..." (MM)

And there is even joking:

> "You know what I really remember that I think is kind of special is Woody [Widmark], our tribal president, he would always—everyone would be like joking about it and he'd like, send out an email it says, 'Is everyone egg-cited?' And he'd have all of these puns about eggs. I still remember 'egg-cited,' because everyone was just, you know, like a little bit high on that." (HD)

In town, the sharing of herring eggs becomes a major "social event." According to Amelia Wilson of Huna Heritage Foundation, when Huna folks learn that their boat (Vernon Hill's the *Shirley N*) is coming back with eggs, "Everyone is so excited. It's like we're just waiting you know, like a kid, just eager. And it's like Christmas, you know. Everyone is so excited, and you get notified and it's like a rush down and...It is really exciting. People are watching [for the boat]" (AN). Then when herring eggs arrive at the dock, everybody drops their current task and makes their way to the boat: "No, no matter what you're doing it's like, 'Oh sorry, gotta go. I'm going to get herring eggs.'" (AN)

In Hoonah, herring eggs are particularly special because they are brought in from afar for everyone:

> "[I]t's so special. It's special. The fact that it has to come from another place and that people have to be kind enough to share. The place where there's a nice big, well, not as big as it was, but they still spawn, so that there's

> enough for that community and for others. I think that has a huge part to do with it." (DW)

The sentiment is similar in Sitka, but because the spawn is centered there, the lure of first fish, the first wild food of the season, is particularly strong:

> "[W]hen I eat it I think of Sitka. It's very unique. We're probably the only place that really gets it—well, used to get it this—I mean, broad spawning grounds and lots of herring eggs." (CM)
>
> "[E]veryone gets really excited along that time. You know, there starts to be a lot of calls. Like, 'Hey, are you getting any?' or yeah. I made an offer to a friend, kind of a work friend, 'Hey I'm going to go to Sitka for this weekend, you know it's herring…' and she jumped right on that. She wanted to come with me." (HD)
>
> "It's the first [fresh food] of the year. It's interesting that even, I think other cultures did that First Fish thing …" (MM)
>
> "When I started getting that stuff [loads of herring eggs], we'd always set up a table … at my folks' house. And we'd always take a couple loads there and they would call their friends and say, 'Come!…' [A]nd it was funny because they would, if they saw my truck there, they'd stop." (MM)

Feasting is a regular feature of bringing home herring eggs. Ken Grant of Hoonah used the word "glee" to describe the comity and convivial animation that pervades the first feast of the year on herring eggs:

> "I think a lot of it was animation of… I see my grandparents and my parents; oh the glee! The happiness, it's just… And we talk about it but we... just celebrated it. We ate as a family…" (KG)

> "It's that huge pile of herring eggs in the center and then everybody would be grabbing… I don't even think we had dishes. We put newspaper out. It's all over the floor." (KG)

Significantly, there was often a spiritual element to this feasting; as a kind of invocation, those consuming herring eggs might speak in Tlingit to their deceased ancestors, who had taught them how to harvest and prepare herring eggs and other values. And when partaking of the initial herring egg meal, Ken Grant remarked, "My dad used to say...all the Elders [in Hoonah used to say]: "*Ch'as x̱át ák.yá x̱waa̱xáa*, 'Was it just me that ate this?' What it means is 'all the ancestors ate with me.'" This query thus serves as a manifest reflection on the alimentary communion and conviviality with ancestral relatives that occurs at traditional first herring egg meals. The consumption of herring eggs thus links consumers not only with living relatives, but also with ancestral ones, and with their regional neighbors, herring populations, and food webs.

After the initial feast on herring eggs, this same table might become a setting for further processing of eggs for distribution, so others too could enjoy the bounty.

> "And I have many, many memories of many years of sitting at my mother's house and my sisters and me sitting around the table and eating fresh herring eggs off of branches, and cutting the branches and bagging them, and distributing them to the rest of the family, and sending them off to Barrow, to Kiana…—in exchange for some caribou…" (LB)

A common distribution pattern for regional centers or cities, it seems, was for an Elder to be identified for primary distribution, say in Anchorage or Seattle, who would, in turn, invite others in her (and often it was a female Elder) family to come and join a herring egg feast, and perhaps receive a secondary distribution of eggs. Tammy Young of Sitka, who used to live in Spokane, explains that every year her family would visit her Aunt Virginia near Seattle to share traditional food and community:

> "We were getting herring eggs and gumboots, clams, all varieties of food from the land from Alaska. We had an aunt that lived in Federal Way just outside of Seattle there, and we would travel from Spokane... Her name was Virginia Bean. And we would travel over there, and it would just be amazing when we'd arrive at her house because people came from a lot of different places in Washington to have a feast. We'd arrive, and her table would be covered with layers of newspaper and water would be boiling on the stove, and there'd already usually be clams and crab on the table. And then all the stories would come out. If we were lucky, somebody from Alaska would actually bring all these delicacies to us and then they would tell us about the days that they went out and how they harvested, and who was with them. And all of these little pieces of information are important, because when you believe that your food feeds your soul, all those people that touched your food, that imbued their love and respect into that food, it is one of the greatest gifts that we give to one another, the love that we feel for each other, the land and the food that we share. The respect that we give to the food is of the utmost importance." (TY)

George Bennett, who now lives in Sitka but used to reside in Anchorage, recalls he would receive a 50-pound box of eggs from his cousin Al Duncan in Sitka and immediately schedule a gathering to consume and distribute them: "Yeah, usually it's so thick you know... and I would give it out to... We had family in Anchorage as well, you know. And we would share, and we would have feast too." (GB)

Tammy Young, a grandmother and STA employee, further characterizes the social benefits of feasting in terms of information flow and keeping people connected to their cultural identity and homeland in Alaska:

> "We would get all kinds of messages. And sometimes they would call us while we would all be gathered. They'd

Herring eggs on hemlock branches being trimmed for distribution outside the Sitka Tribe of Alaska office. (Photo: Bethany Sonsini Goodrich, Sustainable Southeast Partnership)

> call, and we'd put them on speaker phone and they'd talk to us. You know when—it wasn't like we ever asked anyone to send us these things. It was more to the point of a message to us, like 'We sure wish you were home. We miss you around our table.' These were the kinds of things that they talked to us about, that if you were here, you could hear the whales in the bay. So, it was always a form of enticement to remind us that we belong somewhere, and it wasn't where we were currently at. And also, that unconditional love that just naturally happens between family members. They wanted us to feel connected." (TY)

Connectedness can be communicated through the sensuous experience of enjoying key subsistence foods—particularly herring eggs, which are so profoundly tied to time (spring) and place (Sitka)—in alimentary communion with relatives:

> "[I]t's just so important to Elders to you know, smell smoked fish and you know, all these things that they grew up with that... You know, when I was taking care of my mother and grandmother, there was just nothing like when I would make them herring eggs. You know, I mean, they would just light the whole room up; they'd be so happy and dancing in bed. It was just so much fun, yeah. Yeah, Native food just brings out the best in all of us I think." (TY)

By way of a counterfactual to the communion of eating herring eggs together, Tom Gamble remarked: "I don't remember a time where I was eating herring eggs on my own. Have you ever tried to eat herring eggs on your own? It's kind of weird (laughs)" (TG). This was corroborated by FH, a woman from Metlakatla, who recalled receiving eggs from her mother while she was a student in California:

> "My mom sent it down. She Fed-Exed me a box... while I was in grad school down there. Yeah, I was just so happy to get it, you know, I was like 'Alright!' But I realized as I sat down, I was eating it and it didn't taste the same. The whole experience was really weird, and I realized it was because I was eating it by myself. And that was the first time that I'd ever actually eaten this by myself and it was—I didn't really care for it. I mean, you know, you always love that food, but it was just the whole experience [of eating alone] was sort of deflating a little bit." (FH in TG)

Mike Miller commented simply: "I don't think you can get them fresh and not have a dinner right [then, that evening]" (MM). Louise Brady reflected on the intersection of herring egg time and Easter and noted that herring eggs often would be served at Easter as a special dish:

> "[T]hey were fresh [in the spring]. Also, we have family dinners and that's mainly where we have herring eggs because we get so little any more, we want to share. That was really good [referring to dinner] (laughs). We had herring egg salad and then just fresh herring eggs for the Elders." (LB)

Just as Easter is a celebration of resurrection and the renewal of life, the gathering of herring eggs is a stimulus for conviviality, alimentary communion, and the celebration of the spring renewal of life, of family ties and exchange networks, and the beginning of the hunting, fishing, and gathering season. As David Kanosh summed it up: "It's a big social event for my family to be sharing herring eggs with each other." (DK)

Skills, Values, and The Good Life

Skills are essential to successful conduct of subsistence herring egg gathering. The process of enskilment may begin at an early age, as children often accompany their older relatives on egg gathering trips. Here they will learn to sense the fish both directly, through herring's own manifestations, and indirectly through the presence of seabirds, whales, seals, and other predators also looking for herring. They will learn to monitor the coastal zone and see how herring are moving, surfacing, scouting spawning areas, spot-spawning, and finally spawning. They will learn to select the proper kind of Western hemlock, of a certain (young) age, so that the bark and needles do not shed on the eggs, to construct and respectfully deploy sets in the intertidal zone, how to mark them for retrieval, or to conceal them from potential thieves. They will learn to check the sets, how to move them to more productive areas, how to decide when to pull them, and how to grade, trim and process the eggs into coolers or bags or boxes. And then they will learn the arts of preservation (freezing, or occasionally drying) and storage, and they will learn how to give the vast majority of their eggs away to relatives and friends, and the warmth and positive emotions engendered through this sharing, gifting, and reciprocity. Finally, they will learn how status and prestige associated with generosity, giving, and trade helps build social capital, resilience, and extended relational networks.

Learners thus will come to appreciate how herring egg production, distribution skills, and values are part of an integrated knowledge-practice-belief system that serves as a foundation for wellbeing. Such training ultimately stimulates the aspirations, opportunity, dignity, and purpose to realize what anthropologist Edward Fischer (2014) calls *The Good Life.*

> A good life certainly can be happy, but it also involves sacrifice: Amartya Sen [ii] reminds economists of the importance of what he calls "commitment," people's willingness to make counter-preferential choices because of a commitment to an idea or ideal. And such commitment is deeply meaningful for identity, subjective well-being, and one's ability to lead a good life.
>
> A key part of this sacrifice and struggle is the hard work [and enskilment] of becoming a certain sort of person – living up to, and not, certain ideals. Thus, the good life is not a state to be achieved, such as enlightenment, but rather the often arduous journey itself. (http://kingsreview.co.uk/articles/economy-happiness-good-life-interview-edward-f-fischer/)

Developing the key skills and values associated with the good life that bountiful herring brings enables fulfillment not just of the individual but also of the group. The knowledge-practice-belief complex associated with herring production, distribution, and consumption becomes a basis—a cultural model—for aspirations and proper conduct in other domains—in short, for a good life.

This is also true for the skills that underlie herring egg distribution, which also are learned from Elders. Whom do you prioritize in distribution? How do you give? What constitutes appropriate reciprocity in giving, barter, and trade in a complex and changing social-ecological system? How are all the connections embedded in herring egg distributions initiated and maintained? And, finally, what skills and cultural sentiments are involved in receiving the gift of herring eggs, so as to stay connected to kin and homeland.

Vivian Mork (Mork 2019), a recent migrant to Juneau from Sitka, sums up the importance of sharing herring eggs from the receiving perspective:

> "Our herring have huge spiritual, cultural, and social significance, not only to the Tlingit, but to the many other Alaska Native tribes who we share our herring eggs with. Herring eggs are shared with people from as far away as Barrow to as close as our neighbors in Ketchikan. For the Tlingits who've moved away from home, it's our soul food, keeping us connected to one another and to place. If you receive herring eggs from someone, you know you're loved" (VM).

Ultimately, it is this "total social phenomenon" of the gift (Mauss 1967) of herring eggs, the feelings of love, conviviality, renewal, bounty, connectedness, and belonging—of the good life and wellbeing—they engender that continues to motivate the significant levels of Native participation, effort, enskilment, sharing and exchange, adaptation, and resilience that characterize the Sitka Sound herring egg distribution system. In this respect, it is unlike any other subsistence resource in the Southeast Alaska Native portfolio. And, because of this, the Indigenous social-ecological system is considered sacred and vigorously defended by Sitka Natives.

Given the rich multiplicity of cultural beliefs, values, practices, and benefits that underlie the Sitka Sound herring egg cultivation and distribution system, it should be recognized as a cultural treasure, worthy of state, national, and perhaps even international (world heritage) support. Instead, contemporary fisheries management has continued to harm the health of this unique social-ecological system of cultivated abundance, and to put unnecessary stress on the herring egg distribution system that delivers so many benefits to communities throughout Alaska and the Northwest region. As suggested in the concluding chapter, this should be a matter of public policy concern, review, and reform.

CHAPTER 7: CONCLUSIONS AND RECOMMENDATIONS

"If the current management approach continues, we will see the extinction of another cultural tradition, to the detriment of all Alaskans, herring, and the species that rely on herring."

—Rosita Worl, Juneau

Photo: Bethany Sonsini Goodrich, Sustainable Southeast Partnership.

Who will we be when there are no more herring?
Who will feed us?
—Vivian Mork (2019)

Everyone's gonna lose [if we don't take care of the herring].
—Herman Kitka, Sr.
(1997 Board of Fisheries testimony)

This report has provided details on the distribution of herring eggs from Sitka Sound, Alaska. Pacific herring (*yaaw*) are a foundation, cultural keystone, and bellwether species for marine ecosystems and Indigenous culture. When herring arrive to spawn each spring, Sitka Sound comes alive with marine life and becomes the hub for a groundswell of cultural activity, marking the arrival of *yaaw* and the beginning of a new year in the subsistence and marine production cycle. In terms of Native cultural significance, it is among the most important events of the year in Southeast Alaska. The reason for this is the rich complex of production, distribution, consumption, and ceremony that surrounds the customary and traditional subsistence herring egg harvest at its mecca and last stronghold: Sitka Sound.

Summary of Key Attributes of the Sitka Sound Herring Egg Social-Ecological System

Key attributes of the Sitka Sound herring egg system are as follows:

- Production, distribution, and consumption are part of an integrated knowledge-practice-belief complex defined by key cultural values of respect for herring, sharing, barter, trade, and communal feasts and ceremonies. This begins with the blessing of the herring and the herring fishers at Herring Rock (*Yaaw Teiyí*) and continues with the scouting, setting, and monitoring of the spawn and harvest activities, through to every gift or exchange of herring eggs on branches or spawn on kelp that moves this cultural keystone species from Sitka Sound to someone's table. Once distributed, herring eggs are almost invariably eaten with others in a convivial atmosphere of celebration, cultural identity, heritage awareness, and gratitude for the annual "coming alive" of the marine food system.

- Archaeological data, territorial fisheries reports, oral historical and traditional knowledge all support the fact that herring populations were historically much higher and consistent in their returns, abundance, and extensity than currently, and that critical spawning populations have been lost or severely compromised since the mid-twentieth century. If we consider the evidence of the magnitude of herring declines, it is clear that the Southeast herring population today is but a small fraction of what it used to be before industrial herring fishing. Resident stocks, which are said to have complemented the migratory ones (Thornton, et al 2010a) in places like Sitka Sound, have all but disappeared. This historical ecological trend has fisheries management implications in that, currently, Southeast herring stocks continue to be exploited according to a maximum sustainable yield (MSY) model, which is not serving to rebuild the population and has proven ineffective in stemming declines in herring spawning areas throughout the region.

- Even when herring stocks were more plentiful throughout Southeast Alaska, the Sitka Sound herring egg fishery was always a regional one, with Sitka Sound as the mecca for Tlingit participants from as far away as Yakutat, Ketchikan, and the interior and also for neighboring Tsimshian and Haida communities. This created an influx of thousands of Native visitors to Sitka, as documented by nineteenth century Russian America records, and served to develop and support broad sharing, barter, and trade networks. These traditions of participation and widespread exchange continue today. ADF&G has not made precise estimates of the number of non-locals who travel to Sitka to participate in the fishery as recipients or traders, though there is some data on non-local egg harvesters. Findings from this study show that visitors are generally welcomed by locals, who do not try to restrict access to herring eggs, even as quantities of eggs and choice spawning areas have become more limited. On the contrary, local harvesters often assist out-of-towners, with whom they often have social ties, to be successful, that is not to go home empty handed. This is yet another facet of the generosity and social capital associated with the Sitka Sound subsistence herring egg fishery.

- The distribution of eggs is remarkable in its breadth, especially considering the increasingly concentrated spatiotemporal scale that defines the spawning in Sitka Sound. Typically, fewer than 100 harvesters gather all of the 100,000-350,000 pounds of eggs and give 80-90 percent of them away within a week (though some preserved product may be reserved for later distribution) through well-established modes of sharing, barter, and trade, and through trusted networks, both old and new. To be in these distribution networks is to be connected, supported, and secure. The positive values that herring distribution and consumption have produced extend not only to households and communities around the state, but also to goodwill institutions, such as tribal food distribution programs, SEARCH hospital dining services, and the food services of senior centers such as Swan Lake and the Pioneer Home, and those that serve others in need such as the Salvation Army.

- Distribution of herring eggs touches every Native community in Southeast Alaska, and many Native and non-Native communities beyond the region. Very consistently from year to year, herring eggs flow from Seattle to Utqiaġvik (Barrow, the northern tip of Alaska), and occasionally as far away as the Philippines. Alaska's regional centers, such as Juneau, Anchorage, Fairbanks, Bethel, Ketchikan, and Utqiaġvik, are especially important points of distribution, as are proximal Tlingit communities, such as Hoonah, Angoon, and Kake, and proximal Tsimshian and Haida communities, including Metlakatla and Hydaburg. Juneau and Hoonah have especially strong and multi-faceted ties to the Sitka herring distribution system, having sponsored boats to go over to collect eggs through Alaska Native organizations. Angoon and Kake also have one or more boats and crew that regularly travel to Sitka to bring herring eggs back to their communities. While the primary distribution of herring eggs reached at least 41 communities in Alaska and the Northwest region (Washington, Oregon, California, British Columbia, and the Yukon Territory), the secondary distribution of eggs likely reached at least 8-15 additional communities. Estimating precisely how many people receive herring eggs is difficult because primary distributions of eggs are typically accompanied by secondary sharing, and distribution chains may reach thousands of people. Some of these chains contribute to large events, such as *k͟u.éex'* or community feasts, which may be attended by hundreds of people nearly all of whom will consume some herring eggs. Given the massive primary and secondary distribution of herring eggs, it is likely that 5,000 to 10,000 people benefit from them, depending on the level of harvest. In lean harvest years, the number contracts, but many people may still be able to enjoy "a taste" through community events or other distributions. Indeed, modern recipes, such as herring egg salads, effectively stretch limited supplies so more people can enjoy them. Though grateful for any gift of herring eggs, numerous interviewees suggested they would like to have more eggs if they could, and in the case of Elders, like they used to have before the system became so stressed.

- The distribution of subsistence herring eggs often involves cash and has at some level since as far back as the late nineteenth century. This is because Alaska Natives have been well absorbed into the capitalist economy since that time period and face real cash costs for boats, engines, fuel, freezers, shipping, etc. in order to harvest, process, and distribute herring foods. Although commercial transactions of herring eggs are both illegal and rare, customary trade for cash is not uncommon, especially to urban areas, where desired food items for barter may be limited. Alternatively, relatives (typically, but also friends) of egg harvesters may choose to contribute to the production of herring eggs by donating cash to support production (buying fuel, etc.), which is conceptualized differently from trading for cash (a form of balanced reciprocity).

- The acquisition of Sitka Sound herring eggs, be it for food security as "first fish" or heritage or identity reasons, is of sufficient importance that many relatives and friends of Sitka participants establish themselves as part of the production unit. They do this through direct contributions of labor, materials (e.g. hemlock trees), storage capacity (freezer space, etc.), or cash (for fuel, shipping of eggs, etc.). Others have status as recipients, based on their prestigious place in society as kin, Elders, or leaders, or as trade partners. With increasing disruptions to subsistence production, all of these potential recipients, too, are significantly impacted. Instead of getting what they need, or what is customary and traditional, they may get little or nothing at all, and may even be out "fuel money" or other resources they pitched in to support egg gathering. This, in turn, causes sadness, frustration, and even shame for those harvesters who have become respected for their ability to distribute this iconic and nutritious first fish of the season. They reflect not only the loss of herring but of the values and social relationships that constitute Alaska Native culture as one of mutual dependence, trust, and reciprocity. The Nobel-prize winning economist Elinor Ostrom (2015) long ago recognized that such values are foundational to the sustainability of common pool resource regimes that are communally managed, like the customary and traditional herring egg harvest and distribution system.

- The Sitka Sound herring egg distribution network has been not only extensive but resilient and agile. It consistently moves 40,000 to

100,000 pounds or more of herring eggs, by air or by sea, to dozens of communities and thousands of recipients. Even when the subsistence supply of eggs is depressed, as it has been for much of the past twenty years, and especially in the last five years, herring eggs are still distributed widely. The networks that support the herring egg fishery are important to recognize because they are not based on the traditional subsistence management unit: rural residential communities. Rather these multi-local networks are based on customary and traditional regional kinship, barter, and trade lines, which extend great distances, from Arctic Alaska to California, and from the coast to the interior, just as aboriginal kinship and trade networks did. Research designs (such as for harvest surveys) and management measures (for example, to determine the "amount necessary for subsistence") must recognize these customary and traditional characteristics to render suitable regulations.

- With historical overfishing of herring, especially in the mid-twentieth century, spawning stocks have become depleted and increasingly concentrated in fewer areas. In Southeast Alaska, Sitka Sound remains the last stronghold for herring spawning. Thus, the Sitka Sound herring distribution network—absent one in Auke Bay, Ketchikan, or other places where herring used to thrive—becomes even more critical, indeed central, to those in other communities who have come to substitute Sitka herring for their own lost herring spawning opportunities.

- The benefit flows that come with herring, as a cultural keystone species, specialty item, and first-fish of the spring, are considerable. The reappearance and prosperity of herring each year is closely tied to the material, social, and cultural wellbeing of the Sitka Native community, but also other Southeast Alaska Native communities. In this respect, it is the prototype of cultural keystone and critical subsistence species.

- As research supported and cited by the ADF&G, Division of Subsistence clearly shows (Shewmake 2013, cited in Sill and Cunningham 2017; see also Thornton et al 2010a; Thornton and Kitka 2015), successful harvest and distribution of herring eggs relies on both sociocultural factors (knowledge, skills, time, and means) and ecological ones (sufficiency, quality, duration, reliability, and accessibility of spawn). If either set of factors is compromised, the Indigenous social-ecological system of herring egg production and distribution may become vulnerable.

Disturbing Signs for the Wellbeing of the Herring Egg Distribution System

Although the Sitka Sound herring egg distribution system has proven resilient and adaptive, ultimately the health of local and regional herring populations determines its existence and efficacy, and here there are some disturbing signs:

- With fewer areas of spawn, there is increased competition and jostling among harvesters, including from industry-supported boats that are paid to harvest eggs by the sac roe fishing advocacy organization, Southeast Herring Conservation Alliance. These paid harvesters, who may take more than 40 percent of the total subsistence harvest (Holen et al 2011), are further crowding out customary and traditional subsistence users. As a result, the subsistence participation and success of Alaska Native households, the traditional harvesters and distributors of herring eggs, has been harmed significantly, and the damage to subsistence is repeatedly evident with the annual failures of subsistence gatherers to harvest the amount necessary for subsistence (ANS), ostensibly 136,000 to 227,000 pounds, for all but three years since 2005. With no change in management, the harm is certain to continue in the future.

- Participation rates in herring egg harvesting appear to be falling, rather significantly since 1988 (cf. Schroeder and Kookesh 1990:49; Sill and Cunningham 2016), when 24 percent of Native households and 5 percent of non-Native households were reported to be participating. Since 2016, this figure has dipped to fewer than 30 households participating in the harvest, out of some 3,000 households in the community (about 50 percent being Native), effectively less than 10 percent (or about one third of the 1988 percentage of households participating).

The main reason for this decline is the contraction of the herring egg fishery itself. The duration of spawning has become more contracted, as has its extensity and intensity. With increasing frequency in recent years, subsistence harvesters have complained about not being able to get enough quality eggs to meet their subsistence needs. Many have scaled back or given up because of lower yields, higher unreliability of spawn, increased harvest and distribution costs, more theft of sets, and increased competition for the prime setting areas, including from paid egg harvesters supported by the Southeast Herring Conservation Alliance. The higher costs of subsistence herring egg production for distribution are directly related to failures on the part of the state to manage adequately for subsistence uses.

- This stress serves to make the distribution network more vulnerable, less agile, and less reliable for those who depend on it. In short, poor spawning leads to a more fragile distribution network. If participation rates continue to decline, it is likely that the number of new participants in the fishery will correspondingly decline, and fewer new entrants will become enskilled in the unique integrated knowledge-practice-belief system that underlines herring production, distribution, and consumption. This, in turn, would reduce the resilience and wellbeing of communities that have come to depend on the ecosystem service and cultural benefits of herring and herring eggs.

- State managers are largely deaf to Alaska Native concerns. Despite more than twenty years of concentrated testimony and other public responses from STA and tribal and other citizens voicing concerns or opposition to ADF&G and the Alaska Board of Fisheries management of the sac roe and subsistence fishery, the state has not seriously re-evaluated its biomass model, its spatial and temporal prescriptions for the sac roe fishery, or the rationale for even exploiting sac roe as a commercial product. They have made minor adjustments and concessions to subsistence concerns—for example passing an amended version of an STA resolution in 2012 for a subsistence-only area for fishing in the areas STA defined as core subsistence use zones (and slightly enlarging it, though not to the full requested extent, several years later).

In the meantime, Sitka Tribe has begun to pursue other channels, both legal and cultural, to defend its wellbeing and social-ecological interests in herring, including filing for a preliminary injunction against the state in December 2018 to prevent it from opening commercial herring fishing until the Sitka Sound sac roe fishery's impacts on subsistence users can be reviewed. This lawsuit and other public protests, such as that of the Herring Rock Water Protectors, represent the strongest actions taken to date by the Sitka Native community to protect this cultural keystone species. In taking these actions, proponents stress that herring are important not only to Sitka but the whole of the Southeast region through the Native harvest distribution system and the ecosystem services that provide food to other species of value in the marine food web (e.g., salmon, seals, etc.).

Indeed, perhaps no other wild food tradition is honored so actively and defended so consistently as herring egg harvests. Even when Sitka Tlingits were temporarily defeated by the Russians in the Battle of 1804 and had to withdraw to the Chatham end of Peril Strait, they returned the following spring to assert their rights to herring. They have continued to defend herring throughout the colonial and industrialization eras to the present, culminating in the formation of the Herring Rock Protectors group in 2018 as a "call to action for Sitkans to stand up in solidarity with the Kiks.ádi Herring Women and tell the Alaska Board of Fisheries to protect herring and Indigenous subsistence harvesting rights" (see https://vimeo.com/253758747).

According to its founder, Louise Brady, Herring Rock Protectors is modeled on the Standing Rock Lakota social movement to protect their sacred homelands from being desecrated, if not annihilated, by unwanted industrial development, in this case of the sac roe fisheries. She started testifying against overharvesting of sac roe in the 1990s.

> And in 1997, it was pretty heartbreaking to me to sit and watch as our Elders, I think Mark [Jacobs Jr.] was probably one of the first people up, and when that red light went off and they were like, 'I'm sorry, but your time is up.' Mark was in the middle of being very eloquent and talking about how just horrible the herring roe fisheries are. And I really believed that there was no way we could

Four of the seven members of the Alaska Board of Fisheries attended the January, 2018 k̲u.éex' ('invite" ceremony) hosted by the Herring Rock Water Protectors. The ceremony was a conscious attempt to change the constrained discourse of BOF meetings to a broader celebration and understanding of herring as foundation, cultural keystone, and bellwether species. The BOF members—from left Israel Payton, Robert Ruffner, Al Cain, and Fritz Johnson— were invited to speak. (Photo: Bethany Sonsini Goodrich, Sustainable Southeast Partnership)

have portrayed, conveyed the importance, the cultural, the historical relationship that we have with herring within the context of the Board of Fish.

So, I just really felt it was very important for people to understand that it's also a spiritual connection. And I believe that the best way to do that was to have the *k̲u.éex'*, because the *k̲u.éex's* are all about honoring our ancestors and honoring our relationships to not only each other, opposite clan, but to our ancestors and to our stories and to our history. And the history of the Herring Rock is probably one of the most important histories that we have.

> We ended up inviting the Board of Fish, all of the members to come to the *ku.éex'*. And I want to say, four members attended part of it. It was really nice. We have photos. The Elders spoke. Basically, it was time. We weren't limited to the three minutes. So, anyone who wanted to get up and say anything about the importance of the herring and the changes they see, they saw, have seen throughout their lifetime, were able to get up and share that. We also sang the Herring Rock song. We ran almost exactly as we would have run a *ku.éex'* for a memorial. We had dinner. We had fire dishes. We gave names. We had—we gave out blankets and we dedicated the *Yaaw Tinaa*, the herring copper shield.
>
> Sadly, sadly of course, the Board of Fish is going to vote with the money. Sadly, they still thought that traditional ecological knowledge and our special relationship to the herring was not really enough to have them change their minds about changing the quota or whatever. But, in talking with the people who are most involved in the process, who had no idea exactly what we were going into, one of the young ladies afterward said to me that if she had not been a part of planning the *ku.éex'* and making the gifts and taking part in the *ku.éex'*, she never would have used the word 'sacred' in relation to the herring. And I think that rather, for us, for me as a Kiks.shaa, as a Herring Lady, is the bottom line, is that we have this sacred relationship with the herring and responsibility. And that it goes way beyond the monetary value that is seen by the Western world. So that's it, kind of in a nut shell. (LB)

The commitment of the Sitka Tribe and other Native groups has been similarly strong in protecting herring grounds and traditions from being irreparably harmed by management. It is an issue not just of cultural rights and the meaning of subsistence but also of equitable benefit flows, of ecocide, ethnocide, and impoverishment. As Wade Martin sums up: "It's just getting worse every year. Like last year I got 200 pounds

... A few years ago I about threw that [quality and quantity of eggs] back in the water; I just would've considered it junk, but that's just what we've become." (WM)

In light of both ANILCA and the Alaska Constitution, which are strong in their emphasis on sustainable use of the state's wild, renewable resources for the benefit of Alaskans, the current prosecution of the herring sac roe fishery makes little sense. From the point of view of Indigenous subsistence egg harvesters, herring management violates the spirit of both state constitutional guidance on natural resource management and federal subsistence law.

Management Recommendations

The following recommendations are put forth in light of the results of this investigation.

1. Commission a thorough and strategic management evaluation to improve the stewardship of Pacific herring. This management evaluation should recognize the historical and cultural ecology that has served to develop and sustain the model of "cultivated abundance," which frames Indigenous herring management, as well as the unparalleled distributional system and ecocultural benefits that the Indigenous model supports. The first principle of the review should be to do no further harm to herring by suspending practices that are causing stress to herring populations and to subsistence production and distribution. A second principle should be to support Indigenous knowledge and practices that perpetuate a productive and equitable herring egg fishery, due to the high sustainability values and distributional benefits these cultural systems provide. This would improve the condition of herring in the region and mitigate what is termed "ecocultural distress syndrome" (Rapport and Maffi 2010) –a reduced state of biological diversity (i.e., herring and the species that depend on herring) and cultural diversity (Indigenous herring cultures) or health caused by anthropogenic stress (a flawed MSY management model compounded by historical overfishing). It would also

come closer to an ecosystem stewardship model of management which is increasingly advocated by scientists (Chapin, et al 2009; Thornton and Hebert 2015). Finally, it would shift exclusive focus away from commercial export of herring to foreign markets and toward their Maximum Sustainable Benefits (MSB) for Alaskans.

2. Add herring to the forage fish management plan. Maximum sustainable yield (MSY) management on depleted fisheries according to unreliable, spatially uninformed, and unreliable biomass forecasts, combined with incautious harvest models, is a poor way to protect foundation forage fish (Pikitch, et al 2012; Thornton and Hebert 2015; MacCall et al 2018). In Southeast Alaska this has led to closure of all but two management areas for lack of commercial quantities of herring: Sitka Sound and Craig-Klawock. These remaining areas have become less reliable for subsistence harvesters due to overall reduced abundance and significant fluctuations in returns and spawning, which are both unprecedented and unanticipated by ADF&G forecast models. As a result, subsistence herring harvesters and users throughout Southeast Alaska have suffered either extinguishment of their resource (localized schools of spawning herring) or significantly restricted opportunities to harvest herring eggs and diminished returns on their harvest efforts. As the rationale behind STA's (2016) Proposal 209 states:

> Herring (*Clupea pallasii*) fill the exact ecological niche described in the Forage Fish Management Plan. Herring feed on zoo- and phytoplankton and, in turn, are food for seabirds, salmon, and marine mammals. Herring, especially juvenile herring, are of particular importance to salmon and have been shown to constitute up to 62 percent of the diets of Chinook salmon (Canada Department of Oceans and Fisheries, 2013).
>
> Ecologically, herring are indisputably forage fish. Pretending, by omission, that they are not is a

scientific and legal absurdity. Herring are classified as forage fish by most government agencies, including the United States Geological Survey, the National Oceanic and Atmospheric Administration, the U.S. federal government, and the Alaska Department of Fish and Game (Alaska Research Fisheries Bulletin, 2002). Those agencies acknowledge herring to be critical to the Alaskan food web. In recognition of this, the fishing of herring in federal waters is prohibited and they cannot not be retained as bycatch (Magnuson Stevens Act, 1976).

By contrast, the State of Alaska does not classify herring as a forage fish. Adding herring to the Forage Fish Management Plan would not change or close existing fisheries, but it would bring the State of Alaska in line with scientific evidence, federal policy, practical experience, and official designation.

3. Reassess the "amount necessary for subsistence" (ANS). Under current regulation, the amount of herring eggs necessary for subsistence significantly underestimates the amount that would be harvested if there was a reasonable opportunity to pursue customary and traditional patterns of use to satisfy subsistence needs. Basing this ANS estimate on select contemporary harvest estimates of a species that has been in a depleted status for at least a half century, as the Alaska Board of Fisheries has done with a range of resources, leads to a gross underestimate of subsistence needs. This gross underestimate, in turn, results in inadequate management responses to protect subsistence uses.

4. Suspend the industry supported (Sitka Herring Conservation Alliance) herring egg distribution program. The industry program to supplement herring egg distribution through paying

select harvesters to set hemlock trees and harvest large quantities of herring eggs for give-away is not a customary and traditional subsistence activity. Moreover, the industry program is not only crowding out subsistence users from the best remaining spawning areas in Sitka Sound, but also undermining the values, skills, and relationships that constitute the customary and traditional production and distribution network as a cultural system. The industry program eggs are not distributed according to customary and traditional means or protocols, but simply given away to whoever shows up. Quality is also not assured in the industry give-away because the relationship to the recipient is not defined by subsistence values. As a result, the program generates feelings of mistrust and even negative reciprocity (contra Ostrom) among many local Native subsistence egg harvesters. This too is causing potentially irreparable harm to subsistence users because once the traditional distribution networks become stressed, or cease to function due to poor herring spawning or crowding out, it may be impossible to revitalize the knowledge-practice-belief complexes that support them.

5. Institute a robust collaborative research and management regime. To insure that subsistence needs are met in the future, the Alaska Board of Fisheries, ADF&G, STA, and perhaps the Federal Subsistence Regional Advisory Council for Southeast Alaska, should enter into a Memorandum of Agreement, as ADF&G and STA did between 2002-2010. This MOA would direct the partners to work together constructively towards a strategic review and revised guidelines for management of herring ecosystems to ensure the health and status of the population and the subsistence herring egg production and distribution system. Tribal and agency scientists should be co-funded to collaborate in carrying out research to better understand the social-ecological system that involves herring so they can better respond to the needs of all of its constituents, including commercial, recreational, and subsistence users, as well as other species that rely on herring for ecosystem services.

Recommendations for Further Research

Finally, this study has some limitations, due to its scope and timeframe, which could be remedied by further research. First, although a significant sample of herring egg harvesters and receivers of eggs were interviewed, the researchers were not able to engage participants from all communities with strong ties to the Sitka Sound herring distribution system. In particular, additional studies could focus on participation in the production and distribution of herring eggs from Angoon and Kake, which were not visited for this study, although select harvesters from these communities were interviewed. It seems clear that a highly developed tribally-based distribution does not exist in most communities, as it does in Hoonah and used to in Juneau, but only further research would reveal precisely how distribution systems work in other communities.

Second, it is clear that Native distribution of herring in 2018 was significantly impacted by ADF&G's poor biomass estimate (which led to the early closure of commercial fishing when the full quota of herring did not appear). This, in turn, contributed to the poor return on subsistence egg harvesting that occurred after the commercial closure. Therefore, 2018 was not representative of the full nature and range of activities related to the distribution of herring eggs. A multi-year study and survey could remedy this, as it would show how local production and distribution systems adapt to changes in supply of and demand for eggs over time. Changes in the terms of sharing and exchange could be documented more precisely to ascertain how the system copes with stresses versus surpluses.

Third, we recommend more research time and resources to allow for detailed participant observation and documentation of the benefit flows of herring eggs beyond primary distribution. A key conclusion of the study, however, is that secondary sharing, gifting, barter, and trade are customary and traditional, prolific, and profound. A multi-year study with an in-depth focus on following egg distributions and exchanges, as well as tracking their meanings and benefits, would help to mitigate this limitation and the broader bias current ADF&G and other subsistence studies have in tending to emphasize production and consumption over distribution.

It is therefore recommended that further study of the unique and culturally vital Sitka Sound herring egg fishery be undertaken to supplement the findings of this investigation and to support improved management of herring egg uses according to the findings and recommendations of this study. This is in accordance both with the State constitution, which stresses not only sustainable yield of fisheries but also managing them for the *maximum benefit of Alaska's people*, and federal subsistence law, which emphasizes the priority of subsistence uses because they are *essential to Native physical, economic, traditional, and cultural existence.* Nothing exemplifies these benefits and essentials more holistically and integratively than the Sitka Sound herring egg harvest and distribution system.

REFERENCES

Alaska Department of Fish and Game. 2019a. Herring Fisheries in Alaska: Species Information. Alaska Department of Fish and Game, Division of Commercial Fisheries, online document, http://www.adfg.alaska.gov/index.cfm?adfg=her ring.research, (last accessed 25 January 2019).

2019b. Herring Fisheries in Alaska: Species Information. Alaska Department of Fish and Game, Division of Commercial Fisheries, online document, http://www.adfg.alaska.gov/index.cfm?adfg=herring.uses, (last accessed 25 January 2019).

2009a. Herring Fishery Announcement, December 11. Alaska Department of Fish and Game, Division of Commercial Fisheries, Sitka Sound, online document, http://documents.cf1.adfg.state.ak.us/AdfgDocument.po?DOCUMENT, (last accessed 24 May 2010).

2009b. Southeast Alaska Sac Roe Herring Fishery Management Plan. *Regional Information Report No. 1.* J08-09: 1-20.

Alder Jacqueline, B. Campbell, V. Karpouzi, K. Kaschner, and D. Pauly. 2008 *Forage Fish: From Ecosystems to Markets.* Annual Review of Environment and Resources. 2008 33:1, 153-166.

Anderson, G.D. and Harrison, K.D., 2006. Language hotspots: linking language extinction, biodiversity, & human knowledge base. Salem, OR: Living Tongues Institute for Endangered Languages Occasional Papers.

Balée, W. (ed). 1998. *Advances in historical ecology.* Historical Ecology Series. New York: Columbia University Press.

Bavington, D., 2011. Managed annihilation: an unnatural history of the Newfoundland cod collapse. Vancouver: UBC Press.

Berkes, F., J. Colding, and C. Folke, eds. 2000. Rediscovery of traditional ecological knowledge as adaptive manage ment. *Ecological Applica tions.* 10:1251-1262.

Berkes, F. 2003. Alternatives to conventional management: les sons from small-scale fisheries. *Environments* 31: 5-19.

Berkes, F. 2012. *Sacred Ecology*. Routledge, New York.

Bird-David, N. 1999. 'Animism' revisited: personhood, environment, and relational episte mology. *Current Anthropology* 40 (Supplement): 67–92.

Brady, L. 1997. Testimony to the Board of Fisheries in 1997. Accessed 21 December 2018. https://soundcloud.com/herring-rocks.

Brockington, D. 2002. *Fortress conservation: The preservation of Mkomazi Game Reserve, Tanzania*. Bloomington: Indiana University Press.

Brockington, D, R. Duffy and J. Igoe. 2008. *Nature Unbound: Conservation, capitalism and the future of protected areas.* London: Earthscan.

Brown, Evelyn D., Jody Seitz, Brenda L. Norcross, and Henry P. Huntington 2002. Ecology of Herring and Other Forage Fish as Recorded by Resource Users of Prince William Sound and the Outer Kenai Peninsula, Alaska. Electronic document, http://www.adfg.state.ak.us/pubs/afrb/vol9_n2/browv9n2.pdf.

Carls, M.G., S.W. Johnson, M.R. Lindeberg, A.D. Neff, and P.M. Harris.2008. Status review of Pacific herring (*Clu pea pallasii*) in Lynn Canal, Alaska. Juneau, AK: Ted Stevens Marine Research Institute, Alaska Fisheries Science Center, Auke Bay Laboratories.

Carothers, C. and C. Chambers. 2012. Fisheries privatization and the re-making of fishery systems. *Environment and Society* 3(1): .39-59.

Chapin III, F.S., G.P. Kofinas, and C. Folke (Eds). 2009. *Principles of Ecosystem Stewardship: resilience-based natural resource management in a changing world.* Springer.

City and Borough of Sitka. 2018. Resolution No. 2018-02. Accessed January, 2019: https://sitka.legistar.com/LegislationDetail.aspx-?ID=3305030&GUID=BB30BFBC-E47D-4C64-8795-BC-60174631C6&Options=ID|Text|&Search=herring.

Cohen, K. A. 1989. *Wrangell harvest study.* Technical Paper 165. Juneau: Alaska Department of Fish and Game, Division of Subsistence.

Comberti, C., Thornton, T.F., de Echeverria, V.W. and Patterson, T., 2015. Ecosystem services or services to ecosystems? Valuing cultivation and reciprocal relationships between humans and ecosystems. *Global Environmental Change*, 34: 247-262.

Craig, R. 2010. Can Tribal Co-Management of Sitka's Comercial Herring Fishery Protect a Community's Traditional Herring Fishery? In T.F. Thornton, et al, *Documenting and Modeling Herring Spawning Areas within Socio-ecological Systems over Time in the Southeastern Gulf of Alaska.* Final Report, North Pacific Research Board Project #728. Portland, OR: Portland State University. Appendix D.

de Laguna, F. 1960. *The Story of a Tlingit Community: a Problem in the Relationship Between Archeological, Ethnological and Historical Methods.* Smithsonian Institution Bureau of American Ethnology Bulletin 172, Washington, D.C.

de Laguna, F. 1972 *Under Mount Saint Elias: The History and Culture of the Yakutat Tlingit.* Smithsonian Contributions to Anthropology Vol. 7 (3 Parts). Washington, DC: Smithsonian.

Dowie, M., 2011. *Conservation refugees: the hundred-year conflict between global conservation and native peoples.* MIT Press.

Durkheim, E., 1995 [1912]. *The Elementary Forms of Religious Life*. New translation with an Introduction by Karen Fields. New York.

Dyson-Hudson, R and Smith, E.A. 1979. Human Territoriality: An Ecological Reassessment. *American Anthropologist* 80:21-41

Ellanna, L., and G. Sherrod. 1986. Timber management and fish and wildlife utilization in selected southeast Alaska communities: Klawock. Technical Paper 126. Juneau: Alaska Department of Fish and Game, Division of Subsistence.

Emmons, G. T., and F. de Laguna. 1991. *The Tlingit Indians*: Seattle and New York: University of Washington Press and the American Museum of Natural History.

Exxon Valdez Oil Spill Trustee Council. 2011. Pacific Herring. Accessed January, 2019: http://www.evostc.state.ak.us/index.cfm?FA=status.herring.

Federal Subsistence Management Program. 2003. Customary Trade. Federal Subsistence Management Program. Anchorage: Federal Subsistence Board. Accessed December, 2018: http://npshistory.com/brochures/alaska/customary-trade.pdf.

Firman, A.S. and R.G. Bosworth. 1990. *Harvest and use of fish and wildlife resources by residents of Kake, Alaska*, Technical Paper 145. Juneau: Alaska Department of Fish & Game, Division of Subsistence.

Fisher, E.F. 2014. *The Good Life: Aspiration, Dignity, and the Anthropology of Wellbeing*. Stanford University Press.

Garibaldi, A. and N. Turner. 2004. Cultural keystone species: implications for ecological conservation and restoration. *Ecology and Society* (3):1.

George, G., and R.G. Bosworth. 1988. *Use of fish and wildlife by residents of Angoon, Admiralty Island*, Alaska. Technical Paper 159. Juneau: Alaska Department of Fish & Game, Division of Subsistence.

Grinëv, A.V., 2005. *The Tlingit Indians in Russian America, 1741-1867*. Lincoln: University of Nebraska Press.

Hamada, S., 2015. Ainu geographic names and an Indigenous history of the herring in Hokkaido, Japan. *The Canadian Journal of Native Studies*, 35: 43-58.

Hebert, K. 2017. Southeast Alaska herring stock assessment surveys. Alaska Department of Fish & Game, Fishery Data Series No. 17-01. Anchorage.

Holen, D. J. Stariwart, Lemons, T., Ciccone, V. and Turek, M. 2011. The Subsistence Harvest of Herring Spawn in Sitka Alaska 2002-2010. ADF&G Division of Subsistence Technical Paper No. 343. Anchorage. http://www.adfg.alaska.gov/techpap/TP%20 343.pdf.

Illich, Ivan. 1973. *Tools for Conviviality*. New York: Harper and Row.

Ingold, T. 2000. *The Perception of the Environment*: Essays in livelihood, dwelling and skill. London: Routledge.

Jones, R., Rigg, C. and Pinkerton, E., 2017. Strategies for assertion of conservation and local management rights: A Haida Gwaii herring story. *Marine Policy* 80:.154-167

Jones, R. 2007. Application of Haida oral history to Pacific herring management. In *Fishers' Knowledge in Fisheries Science and Management*, ed. N. Haggan, B. Neis and I. G. Baird, 103-118. UNESCO Publishing.

Jones, R., Rigg, C. and Pinkerton, E., 2017. Strategies for assertion of conservation and local management rights: A Haida Gwaii herring story. *Marine Policy*, 80:154-167.

Kitka, H. 1998. Deep Ties to Deep Bay. IN Crisis in the Last Frontier: The Alaskan Subsistence Debate. *Cultural Survival Quarterly*. Fall: 47-48.

Krause, A. 1956. *The Tlingit Indians*. Seattle: University of Washington Press.

Langdon, S. J. 1991 (May). Tlingit and Haida Indian Harvest, Use and Trade of Herring Roe on Kelp in Southeast Alaska. Unpublished manuscript in the author's possession.

Langdon, S.J., 2007. Sustaining a Relationship: Inquiry into the Emergence of a Logic of Engagement with Salmon among the Southern Tlingits. In M.E. Harkin and D.R. Lewis (Eds), *Native Americans and the environment: perspectives on the ecological Indian*, pp.233-273. Lincoln: University of Nebraska Press.

Langdon, S.J. and R. Worl. 1981. Distribution and Exchange of Subsistence Resources in Alaska. Alaska Department of Fish and Game, Division of Subsistence Technical Paper No. 55. Juneau.

Lisianski, U. 1814. *A Voyage Round the World, in three years*, 1803-1806, in the ship Neva. London: Longman, Hurst, Rees, Orme, and Brown.

Lütke, F.B., 1987 [1835]. *A Voyage around the World, 1826-1829*. Vol1: To Russian America and Siberia. Tras. From the French by R. Marshall. Edited by R. A. Pierce. Kingston Ontario: Limestone Press.

MacCall et al. 2018. MacCall, A.D., Francis, T.B., Punt, A.E., Siple, M.C., Armitage, D.R., Cleary, J.S., Dressel, S.C., Jones, R.R., Kitka, H., Lee, L.C. Levin, P.S., McIsaac, J. Okamoto, D.K., Poe, M. Reifenstuhl, S. Schmidt, J.O., Shelton, A.O. Silver, J.J., Thornton T.F., Voss, R. and Woodruff, J. 2018. A heuristic model of socially learned migration behaviour exhibits distinctive spatial and reproductive dynamics. *ICES Journal of Marine Science.*

Marine Stewardship Council. 2017. Can reduction fisheries be sustainable? *Seafood Matters* [blog]. Accessed December 20, 2018. http://blog.msc.org/blog/2017/03/22/reduction-fisheries-sustainable-fish-oil/ .

McClellan, C. 1975. *My Old Peoples Say: An Ethnographic Survey of Southern Yukon Territory*. Ottawa, Ontario: National Museum of Man, Publications in Ethnology No 6 (1&2).

Maffi, L. 2005. Linguistic, cultural, and biological diversity. *Annual. Review of Anthropology* 34: 599-617.

Maffi, L. and Woodley, E., 2010. *Biocultural Diversity Conservation: A Global Sourcebook*. London, UK and Washington DC: Earrthscan.

Mauss, M., 1967. *The Gift: Forms and Functions of Exchange in Arachaic Societies*. I. Cunnison, trans. New York: WW Norton.

Meuret-Woody, H., B. Norcross, and N. Bickford. n.d. "Stock identification of Pacific herring in Sitka Sound, Alaska using Otolith chemical analysis." Manuscript in authors' possession.

Millennium Ecosystem Assessment. 2005. *Millennium Ecosystem Assessment: Ecosystems and human well-being: synthesis*. Washington DC: Island Press.

Mork, Vivian. 2019. Woven with Herring: Sharing Herring is Caring. *Capital City Weekly*, March 19. https://www.juneauempire.com/news/planet-alaska-woven-with-herring/.

Moss et al. 2016. Moss, M.L., Rodrigues, A.T., Speller, C.F. and Yang, D.Y., 2016. The historical ecology of Pacific herring: Tracing Alaska Native use of a forage fish. *Journal of Archaeological Science: Reports*, 8, pp.504-512.

Newton, R. G. and M L. Moss. 2005. *Haa Atxaayi Haa Kusteeyix Sitee, Our Food is our Tlingit Way of Life: Excerpts of Oral Interviews*. USDA Forest Service, Alaska Region, R10-MR-30, Juneau, AK.

Oberg, Kalervo. 1973. *The Social Economy of the Tlingit Indians*. University of Washington Press, Seattle.

Olson, R.L. 1936. Some trading customs of the Chilkat Tlingit. In *Essays in anthropology presented to AL Kroeber*, pp. 211-214. Berkeley: University of California Press.

Pauly, D. 1995. Anecdotes and the shifting baseline syndrome of fisheries. *Trends in Ecology & Evolution* 10(10):430.

Petitgas, P., Reid, D., Planque, B., Nogueira, E., O'Hea, B. and Cotano, U., 2006. The entrainment hypothesis: an explanation for the persistence and innovation in spawning migrations and life cycle spatial patterns. ICES Document CIEM: B07.

Rapport, D. and Maffi, L., 2010. The Dual erosion of Biological and Cultural Diversity: Implications for the health of ecocultural systems. In *Nature and Culture: Rebuilding Lost Connections,* S. Pilgrim and J. Pretty, Eds, pp.103-22. London: Routledge,

Roberts, C. 2007. *The Unnatural History of the Sea*. London: Island Press.

Rounsefell, G. A. 1930. Contribution to the biology of the Pacific herring, *Clupea pallasii*, and the condition of the fishery in Alaska. *Bulletin of the U.S. Bureau of Commer cial Fisheries* 45:227-320.

Rounsefell, G. A. 1931. Fluctuations in the Supply of Herring (*Clupea pallasii*) in Southeastern Alaska. *Bulletin United States Bureau of Fisheries* 47 :15-56. Washington: U.S. GPO.

Salisbury, O. M. 1962. *Quoth the Raven: A little journey into the primitive*. Seattle: Superior Publishing Co.

Schroeder, R. F. and M. Kookesh. 1990. The subsistence harvest of herring eggs in Sitka Sound, 1989. Juneau: *Alaska Department of Fish and Game, Division of Subsistence,* Technical Report No. 173.

Scott, J.C., 1998. *Seeing like a State: How certain schemes to improve the human condition have failed.* Yale University Press.

Sill, L. A., and M. Cunningham. 2017. The subsistence harvest of Pacific Herring spawn in Sitka Sound, Alaska, 2016. Alaska Department of Fish and Game Division of Subsistence, Technical Paper No. 435. Juneau. http://www.adfg.alaska.gov/techpap/TP435.pdf

Smith, A.D.M., C.J. Brown, C.M. Bulman, E.A. Fulton, P. Johnson, I.C. Kaplan, H. Lozano-Montes, D. Anthrony, M. Smith, S. Mackinson, M. Marzloff, L.J. Shannon, Y.J. Shin, J. Ta. 2011. Impacts of fishing low-trophic level species on marine ecosystems. *Science* 333 (6046):1147-1150.

Sitka Tribe of Alaska. 2009. 2009 Annual Report on Manage ment of Customary and Traditional Herring Harvest (Draft). Sitka Tribe of Alaska, Sitka, AK. 14 June 2010. Draft Report on File at STA.

Sitka Tribe of Alaska. 2007 Sitka Tribe of Alaska, Herring Committee. Herring Update: 1-2. March.

Sitka Tribe of Alaska. 2016. Forage fish proposals. Accessed January, 2019: http://www.sitkatribe.org/uploads/cms/Forage%20Fish%20Proposals%202016.pdf.

Sitka Tribe of Alaska. Press release by STA chair Kathy Hope Erickson. December 11, 2018

Suttles, W. 1990. Northwest Coast, Vol. 7. Sturtevant, WC Handbook of North American Indians. Smithsonian Institute, WA.

Swanton, J. R. 1905. Tlingit Method of Collecting Herring-eggs. *American Anthropologist* N.S. 7:172.

Swanton, J.R.H. 1905a. Haida Texts and Myths: Skidegate Dialect. Bureau of American Ethnology. Bulletin, 29. US Government Printing Office.

Swanton, J.R., 1909. Tlingit Myths and Texts. Bureau of American Ethnology Bulletin 39. Washington, DC: Smithsonian Institution. US Government Printing Office.

Thomas, S. 2014. Union Tells Roe Fishermen to Avoid Haida Gwaii. *The Northern View*. March 5, 2014. May 8, 2018. Accessed December 21 2018. https://issuu.com/blackpress/docs/i20140305070533397/9

Thornton, T. F. with contributions by D. McBride, Saurabh Gupta, et al. 2004. *Klondike Gold Rush National Historic Park: Ethnographic Overview and Assessment*. US Department of the Interior, National Park Service.

Thornton, T.F. 2015. The Ideology and Practice of Pacific Herring Cultivation among the Tlingit and Haida. *Human Ecology*, 43(2): 213-223.

Thornton, T.F. 1997. Know Your Place: The organization of Tlingit geographic knowledge. *Ethnology* 36.

Thornton, T. F. 2008. *Being and Place among the Tlingit*. Seattle: University of Washington Press.

Thornton, T.F., M. L. Moss, V. L. Butler, J. Hebert, and F. Funk. 2010a. *Herring Synthesis: Documenting and Modeling Herring Spawning Areas within Socio-ecological Systems over Time in the Southeastern Gulf of Alaska*. Final Re port, North Pacific Research Board Project #728. Port land, OR: Portland State University.

Thornton, T.F., Moss, M.L., Butler, V.L., Hebert, J, and Funk, F. 2010b. Local and Traditional Knowledge and the Historical Ecology of Pacific Herring in Alaska. *Journal of Ecological Anthropology* 14(1):81-88.

Thornton, T.F. and Mamontova, N. 2017. Hunter-Gatherers and Fishing Rights in Alaska and Siberia: Contemporary Governmentalty, Subsistence, and Sustainable Enterprises. In *Hunter Gatherers in a Changing World* , V. Reyes-Garcia and A. Pyhälä, A. (eds). Springer.

Thornton and Hebert J. 2014. Neoliberal and neo-communal herring fisheries in Southeast Alaska: Reframing sustain ability in marine ecosystems. *Marine Policy* 61: 366-375.

Thornton., T.F. and Kitka, H. 2015 An Indigenous Model of a Contested Pacific Herring Fishery in Sitka, Alaska. *In ternational Journal of Applied Geospatial Research*, 6(1): 94-117.

Tikhmenev, P.A., 1978. *A History of the Russian-American Company* (1861-63); reprint, ed. and trans. by Richard A. Pierce and Alton S. Donnelly, Seattle.

Todd, P.A., 2017. *Yaaw (Herring) & Gáax'w (Herring Eggs): The Knowledge Politics of a Traditional Tlingit Subsistence Foodway in Sitka, Alaska.* Doctoral dissertation, University of Cincinnati. Accessed December 2018: https://etd.ohiolink.edu/!etd.send_file?accession=ucin1511867642668127&disposition=inline

United States Bureau of Fisheries. 1916. *Annual report of the Commissioner of Fisheries to the Secretary of Commerce for the fiscal year ended 1915.* Washington DC: US Government Printing Office.

United States Congress. 1912. *Before the Subcommittee of the Committee on Fisheries, United States Senate, Sixty-Second Congress, Second Session on S. 5856, a bill to amend an act for the protection and regulation of the fisheries of Alaska.* Washington DC: Government Printing Office.

Victor-Howe, Anne-Marie. 1992. Subsistence Harvests and Trade of Pacific Herring Spawn on *Macrocystis* Kelp in Hydaburg, Alaska. Alaska Department of Fish and Game, Division of Subsistence Technical Paper No. 225. Juneau.

Walker, B., and D. Salt. 2006. Resilience thinking: sustaining ecosystems and people in a changing world. Island Press, Washington, D.C., USA.

Weil, A. 2011. Herring or Sardines? Accessed on December 21, 2018: https://www.drweil.com/diet-nutrition/nutrition/herring-or-sardines/

Wheeler, P. and Thornton, T.F. 2005. Subsistence research in Alaska: A thirty year retrospective. *Alaska Journal of Anthropology*, 3(1):.69-103.

Wolfe, R.J., B.L. Davis, S. Georgette, and A. Paige. 2000. Sharing, Distribution, and Exchange of Wild Resources: An Annotated Bibliography of Recent Sources. Alaska Department of Fish and Game, Division of Subsistence Technical Paper No. 263. Juneau.

Worm, B., R. Hilborn, J. K. Baum, T.A. Branch, J. S. Collie, C. Costello, M. J. Fogarty, E. A. Fulton, J. A. Hutchings, S. Jennings, O. P. Jenson., H. K. Lotze, P. M. Mace. T. R. McClanahan, C. Minto, S. R. Palumbi, A. M. Parma, D. Ricard, A. A. Rosenberg, R. Watson, D. Zeller. 2009. Rebuilding global fisheries. *Science* 325 (5940): 578–585.

APPENDICES

APPENDIX I

SUMMARY OF ADF&G COMMERCIAL AND SUBSISTENCE HERRING DATA

Year	Forecast Biomass	GHL Quota (tons) Quota (tons)	Sac Roe Harvest (tons)	Roe Percent	ExVessel $ Value (mill.) Escapement	Estimated Subsistence	Participation %HH Harv.(Giving)	Nautical Miles Spawn
1971	NA	750	278	8.3		NA-	NA	9.0
1972	NA	850	603	NA		-	"	14.0
1973	NA	600	537	8.5		-	"	10.0
1974	NA	600	712	12		-	"	10.0
1975	6,400	550	1,484	11		-	"	8.0
1976	7,300	780	795	10.2		-	"	13.0
1977	5,650	0	0	NA		-	"	11.0
1978	4,500	250	238	11		-	"	13.0
1979	20,300	2,000	2,559	9.3	$5.60	-	"	41.0
1980	39,500	4,000	4,445	10.8	2.15	-	"	63.0
1981	27,000	3,000	3,506	11.0	2.38	-	"	60.0
1982	30,000	3,000	4,363	11.7	3.20	-	24% (NA)	40.8
1983	32,850	5,500	5,416	11.1	5.03	-	NA	68.0
1984	30,550	5,000	5,830	11.1	3.73	-	"	65.0
1985	38,500	7,700	7,475	11.3	7.88	-	"	60.5
1986	30,950	5,029	5,443	11.9	7.41	-	"	51.6
1987	24,750	3,600	4,216	9.9	4.40	-20,494	9% (NA)	86.0
1988	46,050	9,200	9,390	9.5	4.17	-	NA	104.0
1989	58,500	11,700	11,831	9.4	1.18	Subsistence	"	65.5
1990	27,200	4,150	3,804	10.6	7.95	-	"	39.1
1991	22,750	3,200	1,838	8.9	0.21	-	"	44.5
1992	23,450	3,356	5,368	9.4	1.37	-	"	72.5
1993	48,500	9,700	10,186	10.7	3.48	-	"	55.3
1994	28,450	4,432	4,887	11.0	3.63	-	"	58.1
1995	19,700	2,609	2,978	11.8	3.93	-	"	37.3
1996	42,265	8,144	8,249	9.6	14.35	127,174	15% (NA)	45.6
1997	54,500	10,900	11,255	11.5	4.73		"	41.0
1998	39,200	6,900	6,787	10.2	1.65	-	"	64.5
1999	43,600	8,476	9,421	10.7	4.91	-	"	59.5
2000	33,365	5,120	4,619	9.9	2.67		"	54.5
2001	52,985	10,597	12,060	11.3	5.79	Harvest	"	61.0
2002	55,209	11,042	9,936	10.9	4.44	*	71% (95%)	42.6
2003	39,378	6,969	6,956	10.7	3.20	278,799	71% (88%)	47.1
2004	53,088	10,618	10,617	10.8	5.16	381,226	60% (93%)	79.8
2005	55,962	11,192	11,520	11.5	6.12		52% (82%)	39.5
2006	52,059	10,412	9,967	10.5	2.64	219,356	55% (91%)	57.4

2007	59,519	11,904	11,571	11.4	5.70	87,211**	48% (89%)	50.2
2008	87,715	14,723	14,386	11.5	8.90	71,936**	41% (73%)	55.3
2009	72,521	14,504	14,776	11.8	12.70	213,712	48% (89%)	65.6
2010	91,467	18,293	17,874	12.5	12.10	154,620	30% (85%)	87.7
2011	97,449	19,490	19,429	13.3	5.17	83,443**	35% (94%)	78.3
2012	144,143	28,829	13,231	11.8	8.50	115,799**	43% (84%)	55.9
2013	76,993	11,549	5,688	13.0	4.44	78,090**	63% (86%)	61.3
2014	81,663	16,333	16,957	12.4	3.05	154,412	68% (88%)	50.0
2015	44,237	8,712	8,756	11.9	2.19	106,998**	66% (57%)	87.9
2016	74,707	14,941	9,769	10.7	2.20	84,554**	37% (75%)	63.3
2017	73,245	14,649	13,923	11.3	4.29	NA**	NA	64.4
2018	55,637	11,128	2,926	11.2	N/A	NA**	NA	33.1
Average 1971-2012	44,212	7,238	6,810	10.8		167,436		49.6

* Pre-1980 Estimated escapement based on either hydroacoustic surveys or applying a conversion of approximately 45.
*1980 to present estimated escapement from most recent age-structured assessment (ASA) model run available as of 2017.
** Years where STA claims subsistence harvest needs were not met according to ADF&G harvest figures. In 2001, STA also reported to ADF&G that their subsistence needs were not met and that they were concerned about impacts resulting from the commercial sac roe fishery (Sill and Lemons 2012)

APPENDIX 2

ADF&G-STA HERRING EGG HARVEST SUPPLEMENT

HARVESTS: HERRING EGGS **HOUSEHOLD ID**

BARTER/EXCHANGE/SHARING

The next set of questions are about sharing herring eggs and the form that sharing takes. We are considering three types of distribution: sharing, in which eggs are gifted with nothing expected in return; barter, in which eggs are exchanged for things other than money; and trade, in which herring eggs are traded for cash.

1. From where have you received herring eggs other than your own harvest?

    relatives

    friends

    STA

    Sealaska

    Southeast Herring Conservation Alliance

    Other: ____________________

    None

...1.a. what are your preferred sources of herring eggs? ____________________

2. Do you SHARE herring eggs? Y N

3. Do you BARTER OR TRADE herring eggs? Y N

If NO to questions 2 AND 3, end this portion of survey. If YES to questions 2 OR 3, continue below.

If YES to question 2 (SHARE)....

......2.a. What is your relationship to the people you SHARE with? ____________________

......2.b. In what communities do they live? ____________________

...2.c. do any of these people contribute directly to your production of herring eggs through any of the following? Y N ?

labor (e.g. help with scouting harvesting or processing)	relationship	____________________
equipment (e.g. boats, trees/branches, grappling hooks)	relationship	____________________
expenses (e.g. fuel money)	relationship	____________________
other: ____________________	relationship	____________________

...2.d. To your knowledge, do these households further distribute the eggs you share to other households? Y N ?

If YES...

.... to where? (circle all that apply) *within their community* *outside of their community* *unknown*

(Continue on following page)

HARVESTS: HERRING EGGS | **HOUSEHOLD ID** |

BARTER/EXCHANGE/SHARING

If YES to question 3 (BARTER OR TRADE)... Y N ?

......3.a. What is your relationship to the people you barter/trade with? ______________________

......3.b. In what communities do they live? ______________________

...3.c. do any of these people contribute directly to your production of herring eggs through any of the following?........................ Y N ?

labor (e.g. help with scouting harvesting or processing) relationship ______________

equipment (e.g. boats, trees/branches, grappling hooks) relationship ______________

expenses (e.g. fuel money) relationship ______________

other: ______________ relationship ______________

......3.d. What items have you received in exchange for herring eggs? ______________________

____

____

...3.e. To your knowledge, do these households further distribute the eggs you barter/trade to other households?........................ Y N ?

If YES...

.... to where? (circle all that apply) *within their community* *outside of their community* *unknown*

4. Are there special people that you always give eggs to regardless of the amount harvested?........................ Y N

If YES...

How are they related to you? ______________________

5. Does your distribution of herring eggs change in a higher harvest year vs a lower harvest year?........................ Y N

If so, how?

i.e. in a high harvest year, do you expand your distribution to new people? What type? In a low harvest year, do you distribute fewer eggs to the same annual people or do you distribute to fewer people? How do you decide who to prioritize? ______________________

____

____

____

____

____

End of the barter/trade additional questions. Make sure the rest of the survey is complete!

APPENDIX 3

INTERVIEW CONSENT FORM

Subsistence Herring Egg Sharing & Distribution: Benefits and Flows

CONSENT FORM

You are invited to participate in a study sponsored by the Sealaska Heritage Institute (SHI), entitled "Subsistence Herring Egg Sharing & Distribution." This research project is being carried out by Dr. Thomas F. Thornton (Oxford University).

The purpose of the study is to document the widespread distribution of herring eggs, especially from Sitka area, to members of the Sitka community and other Native communities in Southeast Alaska and beyond. Herring eggs are a culturally significant and highly seasonal food, harvested in select areas each spring, but widely distributed throughout Southeast via sharing and exchange networks. This harvest and distribution requires significant investment but also creates significant benefits which are currently not fully understood.

Thank you for agreeing to participate in this study. We would like to ask you several questions about how you share, trade and otherwise distribute herring eggs, and the ways in which you receive herring eggs following the harvest.

We would like to record the interview with your consent.

By signing below I agree to the following:

1. I voluntarily consent to provide information to Sealaska Heritage Institute for this project and to allow the interview to be recorded through written notes, photography, tape recording, video-taping or other appropriate means.
2. I agree that the recordings shall be placed with Sealaska Heritage Institute for purposes of future research and preservation. Dr. Thomas Thornton will also maintain a copy of all recordings.
3. I understand that the information that I provide may be used for a variety of purposes including written reports, presentations and publications.
4. I understand that my name may appear in the final products unless I choose to be anonymous.
5. I also understand that the information in the interview may be made available to researchers and/or the general public for use in other projects.

If you have questions about the study, please contact Tom Thornton at University of Oxford (Phone: +44-7503770140, or 907-738-3620; email: Thomas.thornton@ouce.ox.ac.uk) or Chuck Smythe at Sealaska Heritage Institute (907-463-4844, or chuck.smythe@sealaska.com).

____________________________________ _______________

Participant signature *Date*

____

Print Name and Contact details (phone or email)

____ *I would like to receive a copy of my interview tape(s) and/or transcript.*

____ *I would like to receive a copy of the final report.*

APPENDIX 4

LIST OF INTERVIEWEES

Recorded Interviewees	Initials	Community
Aaron Bean	AB	Sitka
Abel Ryan	AR	Metlakatla
Al Duncan	AD	Hoonah
Al Wilson	AW	Sitka
Albin Frederickson	AF	Juneau
Amelia Wilson	AN	Hoonah
Andrew Larson	AL	Hoonah (Sitka)
Breylin Martin	BM	Idaho (Juneau)
Charles Skultka	CS	Sitka (Hydaburg)
Chuck Miller	CM	Sitka
Daphne Wright	DW	Hoonah
David Kanosh	DK	Sitka
David See	DS	Hoonah
Dennis Gray Jr.	DJ	Hoonah
Donald Gregory	DG	Juneau
Doug Chilton	DC	Juneau
George Bennett, Sr.	GB	Sitka (Hoonah)
Gilbert 'Buddy' Hudson	GH	Metlakatla
Harvey Kitka	HK	Sitka
Heather Powell	HP	Hoonah (Sitka)
Helen Dangel	HD	Sitka
Hoonah Senior Center, Glen Johnson et al.	GJ, W, W2, W3, M, RH)=	Hoonah
Jerrod Galanin	JG	Sitka
Jeff Feldpausch	JF	Sitka
Jeff Jackson	JJ	Kake
Ken Grant	KG	Hoonah
Louise Brady	LB	Sitka
Melody Leask	ML	Metlakatla
Mike Miller	MM	Sitka
Mike Baines	MB	Sitka
Mike Smith	MS	Sitka (Kake)
Nels Lawson	NL	Sitka
Anonymous	PE	Sitka
Roberta Littlefield	RL	Sitka
Tammy Young	TY	Sitka
Tom Gamble	TG	Sitka
Vernon Hill	VH	Hoonah
Vivian Mork	VM	Sitka (Wrangell, Juneau)
Wade Marty Martin	WM	Sitka (Hoonah)
Walter Wagner	WW	Metlakatla
Woody (Lawrence) Widmark	WK	Sitka

Note: Additional information was derived from informal interviews and conversations with individuals in Anchorage, Angoon, Juneau, Kake, Ketchikan, Klawock, and Yakutat who did not consent to being identified.

APPENDIX 5

DECEMBER 18, 2018, KCAW NEWS STORY ON STA'S HERRING LAWSUIT DUE TO STRESS ON THEIR SUBSISTENCE EGG PRODUCTION AND DISTRIBUTION SYSTEM.

Jeff Feldpausch [of Sitka Tribe of Alaska] stands in front of bags of hemlock branches ready for distribution to Elders. He noted the bare spots on the branches, illustrating the annual need for subsistence coming up short (Photo/KCAW Emily Kwong).

A tribal government is filing suit against the State of Alaska, alleging mismanagement of the Sitka sac roe herring fishery. The Sitka Tribe of Alaska has retained a major Anchorage law firm that specializes in tribal advocacy and subsistence issues. KCAW's Katherine Rose reports on the build-up to this moment over the future of Sitka's sac roe herring fishery.

For twenty years, tribal leaders have been worried about the health of Sitka's herring. The silvery fish return every spring to spawn and are pursued by commercial fishermen, subsistence harvesters, and marine mammals alike. As a forage fish, they're a cornerstone of the ecosystem.

Herring rock is blessed with water to mark this seasonal moment.

Jessie Johnnie told the story of herring rock to the Board of Fish in 1997 — one of a young Tlingit woman sitting on the rock and lowering her hair into the ocean for the herring to lay their eggs. "All the herring would come to the rock and swim around," she said, "and she would sing lullabies to them."

Herring have cultural, ecological, and economic significance for Sitka. But the message to the Board of Fish back then was that the herring weren't spawning the same way in the same places, and subsistence harvesters were struggling to gather enough roe.

Herman Kitka, testifying at that 1997 meeting, feared for the worst. "If nothing is done," he said, "we will lose the herring stock that is left in Sitka Sound."

In 2018, his son Harvey Kitka went before the Board of Fish to say the same thing: Act now, or potentially lose our herring. Sitka Tribe proposed capping the commercial harvest of herring at 10-percent. But the Board took no action, maintaining a formula that calculates a sliding scale of 12- to 20-percent depending on the size of the biomass.

KCAW's Emily Kwong spoke with Kitka afterwards. He said he wasn't surprised by the Board's decision, but wondered if his father's forecast was coming true.

"It's happening right now, what we were concerned about back then," Kitka said. That was January of this year. In March, the herring fishery opened and ran into trouble. Now, the thing you need to know about the commercial herring fishery is it's driven by processors. They need fish of a certain quality to market their product, largely to Japan and other Asian countries.

Area management biologist Eric Coonradt said the fish this year — most of which were four year olds — were simply too small.

"The quality with which processors needed to market these fish was 125 g or better and 11 percent roe or better and if you look at our forecast, 92 percent of the fish didn't meet that demand," said Coonradt. In other words, they were looking for the biggest and best fish out there, but didn't find enough. The fleet fell over 8,000 tons short of their quota and the commercial fishery closed early for the fourth time in six years. The fishery is driven by the formula and Coonradt noted, it's up to the Board of Fish to change it. "Unless we had a biological concern, we couldn't close this fishery ourselves. What they'd have to do is bring it to the Board of Fish as an emergency petition. That's their option," he said.

Subsistence harvesters didn't have much luck either. Jeff Feldpausch is the Resource Protection Director for Sitka Tribe. His team is bagging hemlock branches covered with eggs for distribution to Elders. He points to bare spots on the branches.

"People don't want trees in their freezer, it's all about putting eggs in their freezers not branches," said Feldpausch. "This is looking grim. This is really grim."

Although the harvest was insufficient for both commercial and subsistence purposes, the state is preparing for next season's fishery under the same model. In December, the Alaska Department of Fish and Game announced they anticipate a spawning mass of 64,000 tons of herring in Sitka Sound, around 9000 more pounds than was originally predicted — and set a commercial quota at 20-percent of that forecast.

This infuriates the Sitka Tribe of Alaska. On Tuesday, December 11, STA filed a lawsuit against the Alaska Department of Fish and Game and the Alaska Board of Fisheries in Superior Court. They're not calling for an all-out closure of the fishery. They're asking for an injunction against ADF&G requiring them to develop a new management plan for the fishery prior to the start of the season next March [2019]. The Tribe also wants the court to find that the actions of the Board of Fish and Alaska Department of Fish and Game are illegal under Alaska law.

In a press release on Friday, STA chair KathyHope Erickson called for protection of the subsistence way of life. "The time is now," she said "to ensure our people have the chance to fulfill their cultural responsibilities which have been interwoven with the herring since time immemorial, and to fill their freezers. We cannot sit by while the State of Alaska shirks its statutory and constitutional duties to citizens. We demand action by the state."

The Tribe has retained the Anchorage law firm of Landye Bennet Blumstein as legal counsel. The State of Alaska has 30 days to reply to the suit.

(Accessed 1/4/19: www.kcaw.org/2018/12/18/sitka-tribe-sues-state-claims-mismanagement-of-herring-fishery/)

APPENDIX 6

ADF&G COMMERCIAL HERRING FISHING MANAGEMENT AREAS

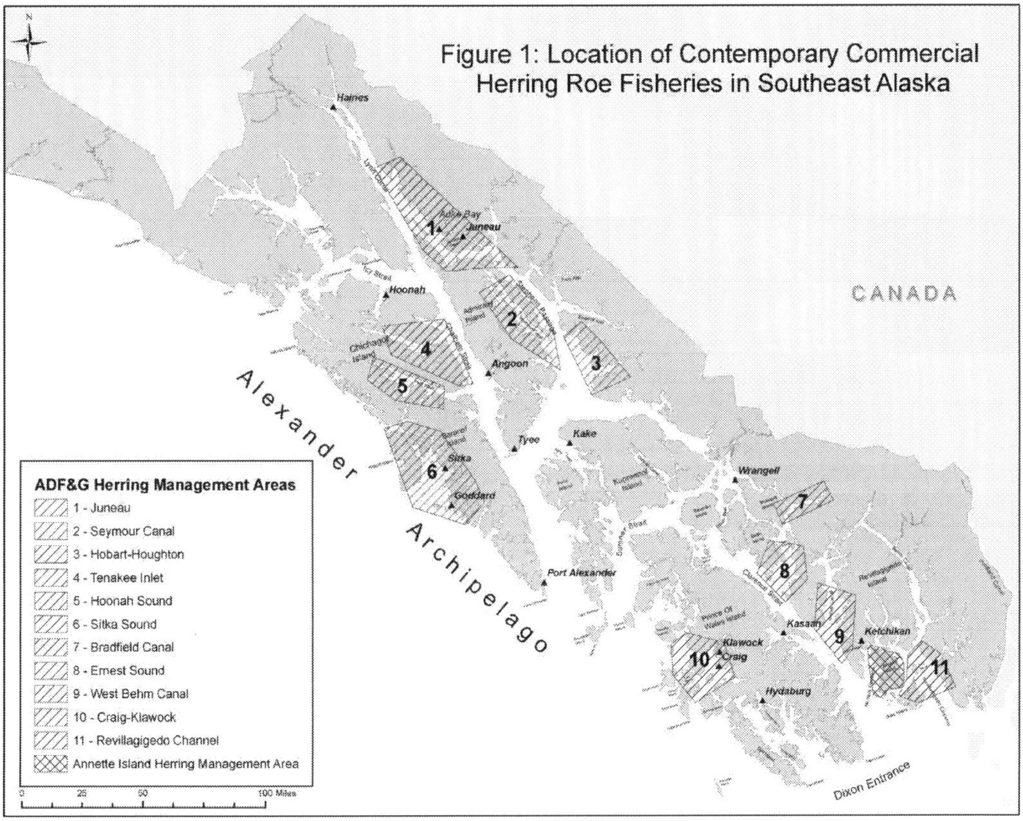

Commercial Herring Management Areas in Southeast Alaska. All Areas are closed for roe fishing as of 2018, but Sitka Sound (#6, sac roe) and Craig-Klawock (#10, spawn on kelp).

CUSTOMARY TRADE
QUESTIONS AND ANSWERS

The Federal Subsistence Board adopted new regulations in January 2003 clarifying the limits of the
caught fish, their parts, and their eggs. The Board's new regulations protect traditional practices of
subsistence fish, but reduce the potential for commercial sale of those fish by prohibiting customary
resale by nonrural individuals. The new regulations allow customary trade transactions between rur
the sale of fish to others for their personal or family consumption only.

What is customary trade?

Customary trade is a term for traditional cash sales of subsistence fish and wildlife.

Do the new regulations defining customary trade of fish apply to other subsistence resources?

No. The regulations passed by the Federal Subsistence Board are only for the cash sales of subsistence fish.

Are sharing and barter affected by these regulations?

No, sharing and barter are not affected by these regulations. Under Title VIII of ANILCA, sharing or barter are protected along with customary trade. These regulations focus only on customary trade. "Sharing" refers to distribution of subsistence foods without receiving something in return; while "barter" is the limited exchange of subsistence fish and wildlife for goods other than cash.

Who can sell subsistence fish?

Only rural Alaska residents are allowed under Federal regulations to exchange subsistence fish for cash.

Can I sell my subsistence fish to anyone?

Yes, with one exception. You may not sell your subsistence fish to any individual or organization required to be licensed as a business under Alaska law.

If you are a rural resident and have an Alaska State business license, a commercial fishing license, and/or a commercial fishing limited entry permit, all customary trade of subsistence fish must be separate from your business or commercial fishing activities.

Can someone who is not a rural resident sell subsistence fish?

No. Only rural Alaska residents are allowed under Federal regulations to sell subsistence fish. Nonrural individuals who purchase subsistence fish must use the fish for personal or family consumption and cannot resell it.

Are there other regulations for selling processed fish?

Yes. The Federal customary trade regulations do not replace State or Federal laws regarding the processing and sale of food for human consumption. All processed foods sold for human consumption must comply with State and Federal food safety laws.

APPENDIX 7

FEDERAL SUBSISTENCE MANAGEMENT BROCHURE ON CUSTOMARY TRADE (FSMP 2003)

What is the definition of processing?

Processing, under State law, includes all activities, which would change the physical condition of the fish. This includes butchering, thermal processing, cooking, dehydrating, freezing, pickling, salting, shucking, or smoking. Some examples of processed fish include dried fish, salted fish, salmon strips, and canning.

Can a rural resident sell processed fish to others?

No. Under Alaska law, if you process fish to sell for human consumption, you must be licensed as a fisheries business. Federal customary trade regulations prohibit the sale of subsistence fish if you are required to be licensed by the State as a fisheries business.

Both State and Federal authorities recognize that rural residents not licensed as fisheries businesses have historically processed and sold subsistence fish in limited/non-commercial quantities. Small-scale sales of processed subsistence fish have not been the focus of law enforcement. The priority for federal enforcement of these provisions will be on large-scale cash sales that are not consistent with customary and traditional practices.

How does the State of Alaska regulate customary trade?

Customary trade is prohibited by State law, unless specifically allowed by the Alaska Board of Fisheries. Current State regulations only allow customary trade of herring roe on kelp in Southeast Alaska.

Can a rural resident sell subsistence fish when the fish are harvested from State waters?

No. Regulations adopted by the Federal Subsistence Board only apply to Federal public lands and waters. This means you can only sell subsistence fish taken from waters under Federal jurisdiction.

Is it legal for the sale of subsistence fish to occur outside of Federal jurisdiction?

It is the opinion of the Federal Subsistence Board that Federal regulations governing customary trade of subsistence fish extend to any customary trade of legally taken subsistence fish, regardless of where the actual cash transaction takes place. However, be aware that State officials may disagree with this interpretation.

Can the regulations defining customary trade be changed?

Yes. If anyone identifies a need to modify or provide for region-specific regulations for customary trade, changes can be proposed through the annual regulatory review cycle.

Many rural Alaskans live off the land, relying on fish, wildlife, and other wild resources.

Alaska Natives have used these resources for food, shelter, clothing, transportation, handicrafts, and trade for thousands of years.

Other residents living in Alaska depend on local harvests as reliable and economic food sources.

Photo courtesy of the Kuskokwim Corporation

The New Regulations for Customary Trade:

1. Provide for continuing customary trade between rural residents. Most customary trade exchanges take place in rural communities and involve small amounts of fish. The new regulations are simpler and eliminate confusing wording found in the old regulations.

2. Prohibit sale of subsistence fish to and purchase by businesses.

3. Require that nonrural purchasers of subsistence fish use the fish for their own or their family's personal consumption. Nonrural purchasers cannot resell subsistence fish.

4. Are consistent with the views expressed by many of the Regional Advisory Councils that the use of fish obtained through customary trade be consistent with the intent of Title VIII of ANILCA.

For more information on customary trade, contact:

Pete Probasco
(800) 478-1456 or (907) 786-3375
pete_probasco@fws.gov

Final Rule: Customary Trade

50 CFR 100.27(c) and 36 CFR 242.27(c)

(11) Transactions Between Rural Residents.
Rural residents may exchange in customary trade subsistence-harvested fish, their parts, or their eggs, legally taken under the regulations in this part, for cash from other rural residents. The Board may recognize regional differences and define customary trade differently for separate regions of the State.

(12) Transactions Between a Rural Resident and Others.
In customary trade, a rural resident may trade fish, their parts, or their eggs, legally taken under the regulations in this part, for cash from individuals other than rural residents if the individual who purchases the fish, their parts, or their eggs uses them for personal or family consumption. If you are not a rural resident, you may not sell fish, their parts, or their eggs taken under the regulations in this part. The Board may recognize regional differences and define customary trade differently for separate regions of the State.

(13) No Sale to, nor Purchase By, Fisheries Businesses.
(i) You may not sell fish, their parts, or their eggs taken under the regulations in this part to any individual, business, or organization required to be licensed as a fisheries business under Alaska Statute, AS 43.75.011 or to any other business as defined under Alaska Statute 43.70.110(1) as part of its business transactions.

(ii) If you are required to be licensed as a fisheries business under Alaska Statute, AS 43.75.011 or are a business as defined under Alaska Statute 43.70.110(1), you may not purchase, receive, or sell fish, their parts, or their eggs taken under the regulations in this part as part of your business transactions.

TTY or modem users may call any Federal agency or program through the Federal Relay Service (800) 877-8339.

Office of Subsistence Management
3601 C Street, Suite 1030
Anchorage, Alaska 99503

Produced by the Statewide Support Division, Office of Subsistence Management

Federal Subsistence Board
Office of Subsistence Management
3601 C Street, Suite #1030
Anchorage, Alaska 99503
(800) 478-1456 or (907) 786-3888
Fax: (907) 786-3898
E-mail: subsistence@fws.gov

www.alaska.fws.gov/asm/home.html

APPENDIX 8

CITY AND BOROUGH OF SITKA RESOLUTION SUPPORTING SITKA TRIBE OF ALASKA

Sponsors: Knox/Bean

CITY AND BOROUGH OF SITKA

RESOLUTION NO. 2018-02

A RESOLUTION BY THE CITY AND BOROUGH OF SITKA SUPPORTING THE SITKA TRIBE OF ALASKA AND THE 2018 BOARD OF FISHERIES DISCUSSION ON SUBSISTENCE HERRING EGG HARVEST IN SITKA SOUND

WHEREAS, the Sitka Tribe of Alaska is the federally recognized tribal government for the Sitka area and is responsible for the health, safety, and culture for its more than 4,400 tribal citizens; and

WHEREAS, the City and Borough of Sitka and the Sitka Tribe of Alaska have entered in to a Memorandum of Understanding "to recognize areas of mutual concern and support, establish a framework for cooperative relations and promote government-to-government communications for the benefit of the community of Sitka as a whole"; and

WHEREAS, the effects of changing ocean conditions brought about by climate change and the significant increase in the humpback whale population on herring stocks are not fully understood nor adequately addressed in the management of the Sitka Sound herring stock; and

WHEREAS, the amount necessary for subsistence (5AAC 01.716 (b)) for the Sitka Sound subsistence herring egg harvest has not been met in five of the last seven years (2010-2016); and

WHEREAS, it is important to recognize the cultural and traditional ecological knowledge in the ongoing discussions and management of the Sitka Sound and all Southeast Alaska herring stocks; and

WHEREAS, the Sitka Tribe of Alaska has submitted three Board of Fisheries proposals that are aimed at conserving the Sitka Sound herring stock and protecting a viable subsistence harvest of herring eggs in Sitka Sound; and

WHEREAS, the Alaska Department of Fish & Game's local Advisory Committee (AC) recognized the need for herring conservation and the concerns of subsistence harvesters; and

WHEREAS, the AC unanimously supported an amended version of Board of Fisheries Proposal 99 to change the Sitka Guideline Harvest Level (GHL) Rate, and the Sitka Tribe of Alaska finds this a reasonable compromise; and

WHEREAS, the Sitka Tribe of Alaska's 2018 Board of Fisheries proposal 106 would close waters adjacent to the core subsistence herring egg harvest area and contiguous with existing waters that are closed to commercial herring fishing for the purposes of establishing a herring conservation zone and protecting the subsistence herring egg harvest; and

WHEREAS, the Sitka Tribe of Alaska opposes 2018 Board of Fisheries Proposals 94 and 104 that are aimed at rescinding existing regulations, which are in place to protect the subsistence harvest of herring eggs in Sitka Sound and to provide a reasonable opportunity for the amount necessary for subsistence to be met.

NOW, THEREFORE, BE IT RESOLVED that the Assembly of the City and Borough of Sitka hereby shares the Sitka Tribe of Alaska's concerns about conserving the Sitka Sound herring stock and protecting the subsistence harvest of herring eggs in Sitka Sound.

PASSED, APPROVED, AND ADOPTED by the Assembly of the City and Borough of Sitka, Alaska this 9th day of January, 2018.

____
Matthew Hunter, Mayor

ATTEST:

____
Sara Peterson, CMC
Municipal Clerk

1st and final reading 1/9/2018

ABOUT THE AUTHOR

Dr. Thomas Thornton, currently Dean of Arts and Sciences and Vice-Provost for Research and Sponsored Programs at the University of Alaska Southeast, has carried out human ecological studies in Alaska and elsewhere for more than 30 years. His book, *Haa Léelk'w Hás Aaní Saax'ú: Our Grandparents' Names on the Land*, which was published by Sealaska Heritage Institute with the University of Washington Press, won a 2012 Alaska Historical Association award. In previous Alaska research he has collaborated with numerous Southeast Alaska tribes and organizations, including Sitka Tribe of Alaska and the Southeast Native Subsistence Commission. While carrying out this study, he served as an associate professor and senior research fellow at University of Oxford's Environmental Change Institute, School of Geography and the Environment, United Kingdom.

Made in United States
Troutdale, OR
11/03/2026

56844357R00155